# NO
# PLANET
# B

# NO
# PLANET
# B

## A *TEEN VOGUE* GUIDE TO CLIMATE JUSTICE

### EDITED BY LUCY DIAVOLO

Haymarket Books
Chicago, Illinois

Published in 2021 by
Haymarket Books
P.O. Box 180165
Chicago, IL 60618
773-583-7884
www.haymarketbooks.org
info@haymarketbooks.org

ISBN: 978-1-64259-259-7

Distributed to the trade in the US through Consortium Book Sales and Distribution (www.cbsd.com) and internationally through Ingram Publisher Services International (www.ingramcontent.com).

This book was published with the generous support of Lannan Foundation and Wallace Action Fund.

Special discounts are available for bulk purchases by organizations and institutions. Please email orders@haymarketbooks.org for more information.

Cover design by Liz Coulbourn.

Printed in Canada by union labor.

Library of Congress Cataloging-in-Publication data is available.

10 9 8 7 6 5 4 3 2 1

# PUBLISHER'S NOTE

For the most part, the voices included in this collection come from activists and artists ages ten to twenty-five. The ages of the people noted in individual chapters and the facts reported are accurate as of the time of publication. Sources of quoted material can be found online.

# Contents

# SECTION TWO: ACTIVISM

## SECTION THREE: INTERSECTIONALITY

# FOREWORD

## LINDSAY PEOPLES WAGNER

*May 2020*

When I started at *Teen Vogue* as an intern at the age of 17, there was almost no mention of climate change and sustainability in any print or online issues. But when I came back as editor in chief a decade later in 2018, it was clear that climate change was an urgent issue that required thorough coverage, especially for our young, engaged readers.

A common misconception I often hear when older people talk about young people's passion for making change is that they're naive and alarmist or that climate activism is over-the-top and distracting. What those older people fail to realize is that for young people, it's tough to consider what their future looks like if the future of the planet is uncertain. If we destroy our planet, we have nowhere else to go. There is no Planet B.

The evidence of climate change is now too copious for even the biggest skeptics to deny with any credibility. We have compelling depictions of how our planet would deteriorate if sea levels continue to rise. We've already seen devastating increases in unpredictable weather patterns like droughts, which in turn create situations like famine. And all the while, ecosystems and species are still depleting rapidly. Simply put, as the planet gets hotter and we stand idly by, we are designing our destiny of fatality.

Too often, it takes things getting worse for things to get better. And too often, our broader culture can only see how bad things are getting through a white gaze. Back in 2016, when Mari Copeny was only twelve years old, she became an activist on behalf of Flint, Michigan, because of the discolored and undrinkable water, an environmental health crisis created by the city's power brokers. Thousands of children were exposed to lead and at risk for developing severe long-term health issues. And that same year, Native American youth stood tall against the Dakota Access Pipeline that would endanger sacred sites and water supplies.

More recently, voices like Greta Thunberg, Jamie Margolin, Xiye Batista, and more have spoken truth to power and demanded justice for the climate. It wasn't until this recent surge of activism in 2019—after weekly school strikes, student-led global days of action demanding action and accountability, and the largest gathering of youth protestors in history—that the world collectively started to take the issue seriously. In the process, they helped reframe the public's understanding of the environment as a justice issue. Climate activist Elsa Mengistu said it best: "If we don't work on climate justice, then we can't work on any kind of justice. Protecting the planet is also about protecting the people on it."

The time for action is now. Young people have shown they're ready to lead the fight against climate change. Even as they do the necessary grassroots work in every corner of our planet, they continue to insist on global action.

To have an equal society and livable planet for all people, we desperately need to understand how climate destruction impacts people from all walks of life. I hope that this book embodies *Teen Vogue*'s motto of making young people feel seen and heard all over the world. I hope that it forces their parents, communities, loved ones, friends, and—most importantly—those in power to see that the health of our planet depends on how quickly and drastically we change our behaviors. I hope it forces them all to respond.

Young people are no longer the future, asking for permission to make changes when it's their turn. Young people are the right now; they are the present actively changing our future for the better on their own.

# INTRODUCTION

# How Climate Justice Became a Pillar of Intersectional Youth Activism

## LUCY DIAVOLO

*May 2020*

y first day on the job as a political news editor at *Teen Vogue* was February 20, 2018—six days after a shooting at Marjory Stoneman Douglas High School in Parkland, Florida, shocked the world. It was an intense time to begin covering politics for a teenage audience, full of heartbreaking stories and gut-wrenching details that were part of a larger, horrible picture about the nature of gun violence in this country.

1

Through the ensuing weeks and months of coverage of an issue that has stalked entire generations now, what kept me moored amid the churning seas of official inaction and policy debate was the rock-solid anchor provided by youth activists from Marjory Stoneman and the young people in communities all across the country they partnered with to expand the public's understanding of gun violence. It was a trial by fire for me but was emblematic of how youth activism and organizing has always been the lifeblood of *Teen Vogue*'s political coverage.

By the time the Parkland anniversary rolled around in 2019, our political conversation had shifted. Activists trying to prevent gun violence were still hard at work, but the media's focus had largely turned to a presidential election getting underway and the continued broadcasts of the narcissistic bluster coming out of the White House. At *Teen Vogue*, as we continued our mission of documenting youth activism, that still meant publishing stories about gun violence, but it also meant the chance to cover other kinds of activism more in-depth.

The climate crisis had always been a key plank in our coverage thanks to Alli Maloney, *Teen Vogue*'s first-ever digital politics editor. This was evident in the December 2018 editorial package, *Plastic Planet*, which covered the myriad ways plastic is destroying the world and much of which is included in this book. Thanks to Alli's work, the climate was already a priority for us by the time we covered Greta Thunberg for the first time ever that same month, December 2018.

Little did I know then that when I asked one of my most reliable newswriters, Emily Bloch, to cover a fiery speech at a United Nations climate change summit, the young woman who schooled world leaders would become the focal point of a resurgent climate justice movement and that I'd get to interview her for a special issue cover story.

Luckily for all of us, the movement is much bigger than Greta. Alongside senior politics editor Allegra Kirkland, executive editor Samhita Mukhopadhyay, and editor in chief Lindsay Peoples Wagner (who made our cover shoot with Greta happen on a wing and a prayer), I've watched as youth-led organizations completely reset the urgency and often the stakes of our conversations about the climate crisis.

While only Greta was named *Time's* 2019 Person of the Year, the new publicity for a movement all around her was very real, and in our efforts to cover it we've tried to be as representative as we can of its expansiveness. To re-create this for this book, I've organized the *Teen Vogue* pieces reproduced here into three sections: reporting, activism, and intersectionality.

The first is a section of reporting intended to put the science first—not just the numbers in isolation but spelling out what it all means. Knowledge is a form of power, and equipping the public with important information gives the people the power to respond to the world around them. For our readers, that means assuming that someone is in the know but never shaming someone who might not be. It means expanding, contextualizing, and connecting information and delivering it in a way that makes people want to keep reading and to share what they learn. Alli captures the ethical imperatives that drive this facts-first mode of operation, laying out how we've followed the lead of youth activists, organizers, and journalists.

The second section is about activism, the engine that has propelled our climate coverage forward. For us, journalism is often a conversation with the youth activists and organizers—not just giving them solid reporting but following their lead on what stories matter and how we should be discussing them. Allegra offers her perspective on documenting the moment 2019 presented and how it fits into the larger arc of climate activism, dating back to the 1970s and beyond.

The third and final section is not focused on an activity like reporting or activism but on Kimberlé Crenshaw's concept of intersectionality. Originally coined to describe discrimination in legal settings, the term's use (and, at times, misuse) has expanded and has an essential application to the climate justice movement. Aspects of our world like race, class, and gender intersect with the climate crisis as the people already on the short end of the stick see the water rising around them. In her introduction, Samhita examines how activists understand both the ways climate injustice operates along existing axes of oppression like environmental racism and how the movement has come to center intersectionality in its perspectives, leadership, and priorities.

Together, these three aspects of the climate justice movement—reporting, activism, and intersectionality—form the guideposts for this collection just as they have guided and continue to guide our coverage day in and day out. Accuracy, advocacy, and equity are foundational pillars to any movement seeking justice, and the climate justice movement of these last several years has embodied them all.

The fact that young activists have centered these same concepts in their work offers a ready-made explanation for how this surge in climate organizing has become a pillar of the modern Left. We cannot afford to lose sight of this moment or lose the energy it has created, as every passing second sends us further into ecological ruin and planetary devastation.

We can trust that the ideas, individuals, and organizations chronicled here will continue to be vital to these discussions as we further attempt to avert, mitigate, and respond to the crisis. Even before the high-water mark moments of 2019, the tide of energy around climate justice was rising. We cannot let it simply roll back out.

I think often of the young Sunrise Movement organizer who joined me for a panel discussion at the Socialism 2019 conference in Chicago. At the end of our rousing discussion, she led a room of comrades in a song that went, "The oceans, they are rising, and so are we."

I hear that chorus of voices in my head to this day, as it continues to be one of the highest honors of my life to document the people rising to this challenge.

# SECTION ONE: REPORTING

# Putting Science, Data, and Facts First in the Fight for Climate Justice

**ALLI MALONEY**

*May 2020*

*T*een Vogue launched its news and politics section two months after the inauguration of a president who rejects and regularly misconstrues climate science. Including environmental coverage in our work was out of necessity: the impact of the climate crisis presents pressing questions concerning cause and effect, and young people were asking the big ones.

The vertical took shape as a commitment to youth struggle in a time of truth-denial against a backdrop of an obscured, erased, and politicized history. In 2017, as today, teens righteously raged against inherited conditions and those eager to dismiss their concerns about rising temperatures and sea levels, catastrophic weather events, and accelerating displacement of frontline communities around the world. The proof was on every continent and confirmed by experts for decades, yet deception and denial remained as calls for drastic measures from scientists and concerned global citizens grew louder.

As deregulation began to rapidly unfold under a new administration sympathetic to oil, gas, mining, and coal industries, the certitude and urgency of the climate justice movement grew. As a convergence of international peers began showing up for one another to share their stories, online advocates expanded their efforts to become on-the-ground organizers. A generation born into catastrophe became a knowledgeable collective fighting for justice and a stable future. All life, they explain, depends on this pursuit—and science backs them up.

Role models emerged. Around this time, water protectors camped out on the Standing Rock Reservation were holding firm in their resistance against a pipeline imposed on Native territory, a battle with implications for everyone who wants to stop the fossil fuel industry. They unapologetically commanded accurate coverage on the destructive impacts of pipelines like the one under construction on sacred lands. Working with elders and allies, young Indigenous revolutionaries exhibited social media and public relations savvy as they drew lines for the public to connect brutal extractive practices, accelerating climate change, and colonialism and white supremacy. In the process, they demonstrated how the empirical evidence of climate science is inextricable from the lived realities of social science, illustrating how we cannot survive on Earth without embodying our respect for this planet in our actions.

And when a movement or message centers science, coverage must do the same. Actively rejecting misinformation is a key undertaking of climate justice advocates and writers and, as a result, became our political approach. To meet our core readership where they're at, we built the lens of our coverage around the one through which they showed us the world.

Their concerns manifested in self-schooling to inform resistance tactics and clarify messaging (and inspire clever protest signage, too). More activists and writers came to us with pitches for reported features and op-eds about the issues. This informed peer-to-peer education attracted an all-ages audience eager for reality in an existentially exhausting era of all-too-eager deceit and corporate spin.

We work to match the seemingly boundless energy of everyone who hears "fake news" about the climate disaster and calls bullshit. To refute the denialists' noise and denounce leaders' inaction, we've studied the subject matter and sought out narratives that give a face, a name, and a story to the images that take shape in data. On the ground in their communities and across the connections of a borderless online network, young organizers have stepped up, uniting peers to strike while citing the statistics and accounts that reveal the real circumstances of this crisis. In tandem, we've strived to recognize and analyze the radical and mainstream environmental movements and figureheads that came before our moment to learn from a past we can't allow ourselves to be doomed to repeat.

This enlightened generation recognizes that humanity's behavior has built up to the cataclysmic human moment in which the activists find themselves. They're eager to interrogate the past, present, and future, and they accept no less than the facts, holding themselves and others accountable. Engaging with the world necessitates confronting urgent problems with curiosity. Whether it's from a concerned older friend or sibling, in headlines about the extinction of a beloved species, or through the bluster of a blowhard uncle repeating the president's lies, hearing "climate change" again and again only sets the stage. The difference between passivity and activity starts with taking the time to get educated, and the work we've reproduced in this section of the book serves that mission.

The realities of the climate crisis are the reason people stand up to stop it. We can see this in the eagerness so many exhibit in understanding these truths. Reliable reporting that makes sense of nebulous forces like the oil industry, energy practices, and plastic pollution has proven essential to a well-rounded understanding of the hotly, wrongly contested complexities of climate disaster.

People in this movement take risks to accurately represent the world. More than one concerned-global-citizen-turned-journalist in this chapter traveled to the Arctic to report from the front line fact-forward accounts that offer teen perspectives and platform the experiences of disproportionately impacted peoples. Centering facts while shifting the lens to prioritize oft-neglected perspectives and redefine expertise is, therefore, a tenant of *Teen Vogue*'s climate coverage. When South Dakota's Rosebud, Pine Ridge, and Cheyenne River Indian reservations were inundated by dire floods caused by unprecedented snowfall and powerful storms, a Dakota/Lakota Sioux writer bore witness and told young readers what she saw firsthand, shaping alarming statistics into a rousing call to action. Reporting on why recycling—a process purported to combat plastic waste—is a broken system, our writer called on a teen organizer from Utah to explain the nuance of its specific laws.

If parents, politicians, and education systems won't (or can't) acknowledge comprehensive realities, someone has to. The movement journalists we've worked with have traveled to frontline communities, interviewed experts and the impacted, dissected research to make nuanced data digestible, and—when all the pieces fall into place—gone back and forth with *Teen Vogue*'s rigorous fact-checkers to establish ironclad veracity for their reporting. In doing so, they are taking the reins of the discourse through tactical activism. To endeavor for climate justice, telling the truth is a crucial element of the practice, a necessary tenet of all-too-necessary work, a challenge accepted. Climate justice activists are leading a conversation that those of us drafting history are humbled to amplify and join.

# What Is Climate Change?

## EMILY HERNANDEZ

*July 17, 2018*

I t's getting hot in here—literally. According to the National Aeronautics and Space Administration (NASA), the global temperature has risen by 1.8 degrees Fahrenheit since 1880, with most of the warming taking place during the last thirty-five years. These increases, known as global warming, have led to major changes to the earth's climate. Unfortunately, these changes are not in the planet's best interest, as with warmer temperatures come a variety of negative environmental consequences, including more frequent and stronger storms, loss of natural habitats, and increased instances of drought and flooding. Unfortunately, one of climate change's main causes—the burning of fossil fuels—is tied to many nations' economic structures, including that of the United States. In other words, what helps to make climate change such a uniquely menacing threat is how the issue is often politicized to the point of generating confusion where there should only be the general understanding associated with scientific facts.

Since U.S. President Donald Trump has denied climate change, moved to deregulate coal, and tapped climate change denier Myron Ebell to lead the Environmental Protection Agency (EPA) transition team, it is up to every individual in the country to educate themselves on the reality that is climate change: it's been affected greatly by human activity and has the capacity to negatively impact future generations.

The climate can't afford any more denial. Here, we unpack the basics of climate change, along with some of the most commonly heard phrases about the process, so you can arm yourself with the facts.

### What is climate change?

The process known as climate change begins with the sun. The sun emits solar energy, which is in turn partly absorbed by the earth's surface and atmosphere and partly reflected back into space. Climate change refers to the environmental phenomenon through which this solar energy, which is responsible for heating the planet, is prevented from being reflected back into space because of the presence of greenhouse gases in the atmosphere. The planet, therefore, grows hotter and hotter, which leads to a whole host of other environmental changes.

### What evidence do we have that show climate change occurring?

The evidence that shows the climate to be changing is abundant. From global temperature increases to warmer ocean waters to shrinking ice sheets to more-frequent severe weather events, the signs are everywhere: the earth's climate is becoming increasingly destabilized. In fact, scientists have been measuring levels of carbon gas in the atmosphere—a key indicator of global warming, and, therefore, climate change—since at least the late 1950s.

**What causes climate change?**

Climate change is a result of a process known as the greenhouse effect. Since the late nineteenth century, we've known that greenhouse gases warm the earth by trapping heat in the atmosphere. Imagine the atmosphere as a greenhouse encircling the planet: the same way greenhouses are used to grow plants in the winter or in colder regions, the greenhouse effect allows us to live our lives on Earth in comfort. The sun radiates solar energy, which travels into Earth's atmosphere, heating us up. Earth, in turn, emits some of the energy back into space, but some of it is blocked from getting that far by the greenhouse gases in the atmosphere. Greenhouse gases, namely carbon dioxide and methane, trap the heat that's trying to escape back into space. Without these gases, Earth would be freezing and uninhabitable for us. Too much of anything is a bad thing, and an overabundance of greenhouse gases leads to an atmosphere that traps too much heat.

The major player consistently talked about is carbon dioxide, $CO_2$. When we burn fossil fuels, we're emitting carbon dioxide up into the atmosphere where it will stay and trap heat for decades. We use oil to fuel our transportation needs and coal and natural gas to supply our electricity. The more fossil fuels we use, the more carbon dioxide goes into the atmosphere, making the greenhouse effect stronger, warming up the planet more and more.

**What are the effects of climate change?**

Since 1880, we've experienced an increase of about 1.8 degrees Fahrenheit in the global temperature. This change has resulted in melting sea ice, rising sea levels, and more extreme weather events. The hotter it gets, the worse it will be.

## What's the difference between global warming and climate change?

"Global warming" and "climate change" are often used interchangeably as synonymous terms. However, there are specific differences in their definitions. According to NASA, global warming refers to an increase in average surface temperature. Climate change is defined by NASA as "a long-term change in the earth's climate, or of a region on Earth" and refers to all the climate-related changes that occur due to greenhouse gas emissions. This not only includes increasing temperature, but also extreme weather events, rising sea levels, and melting glaciers, for example.

## What if we stopped using fossil fuels right now?

If humans were to stop using fossil fuels, climate change would still continue, and the earth's temperature would keep rising. It all has to do with the thermal inertia manifested by the retaining of heat in Earth's oceans. According to NASA, "Inertia is the tendency of an object to resist a change in its current state." Keeping that definition in mind, imagine you're trying to boil a pot of water on a stove. You put the pot on the burner, turn on the stove, and wait. The water takes a bit of time to heat up and eventually start boiling, but once it's hot, it stays hot for a while—even if you take it off the heat. In that same way, our oceans take time to heat up, but once they're hot, they stay hot. At this point, we are committed to a level of climate change that is largely irreversible.

Despite the disheartening effect of that last sentence, the more carbon dioxide we continue to emit, the worse those effects will become.

### What does President Trump think about climate change?

Trump has famously called climate change a "hoax" invented by China. (Liu Zhenmin, China's vice foreign minister, denied this.) Since taking office, Trump has translated his climate change denial into American policy—most notably, pulling the U.S. out of the Paris Agreement and moving to dismantle the Clean Power Plan.

For all of Trump's verbal and realized antagonism toward the reality of climate change, here's the truth: there is no scientific debate about climate change. A 2016 analysis supported the often-cited statistic that 97 percent of scientists, experts in their fields, agree that climate change is being sped up by human actions. In politics and big business, the reality of climate change means new regulations, policies that restrict actions, and adaptation. The fossil fuel industry seemingly has a vested interest in making sure none of that comes to pass.

### Are there any potential solutions for climate change?

Though it may seem hopeless, we still have options. In your everyday life, try to minimize your own carbon footprint. The Nature Conservancy offers a free online calculator to show you how much carbon dioxide your everyday activities emit. Quick changes? Ride your bike or walk when you can. Carpool if a car is the only option, or use public transportation. Consider limiting consumption of red meat, or at least commit to meatless Mondays: about half of the greenhouse gas emissions from agriculture come from livestock, and the worst offender is beef production.

Most importantly, keep educating yourself on the subject, and get loud about it. NASA has a great website that compiles the evidence, causes, and effects with scientific evidence. If you want the government to know you're serious about the climate, go and make your voice heard. If taking on the government seems like a lot to handle, start with your local or state representatives. Write to or call your congressperson, letting them know your opinion on the issue.

# The Climate Disaster We Fear
# Is Already Happening

RUTH HOPKINS

*April 22, 2019*

few weeks ago, I traveled through the Rosebud, Pine Ridge, and Cheyenne River Indian reservations in South Dakota, home to Tetonwan Lakota of the Oceti Sakowin, or Great Sioux Nation. It's difficult to describe the widespread devastation I witnessed.

As I write this, I am snowed in on the Lake Traverse Reservation. This winter has been one for the record books, with unprecedented snowfall and powerful storms the Weather Channel calls "bomb cyclones" or "snow hurricanes." This season the area has received the most snowfall on record. I've lived in the Northern Plains most of my life, so I'm not one to balk at a few snow drifts, or even a few days of 40 degrees below, with the wind-chill factor.

But this year is something else entirely—and the devastation on the abovementioned reservations was a result of snow, storm intensity, and frequency.

These storms, and the subsequent infrastructure disasters, have been hard on Native communities that are already impoverished and lacking infrastructure. Schools and roads have been closed for a week or more at a time. Snow drifts have been higher than the plow blades. After the snow starts to melt, areas already prone to flooding have nowhere for the water to go. I saw a number of roads that were under water and impassable; animals stranded on high ground; people stuck in their homes for weeks. Roadways, where flooding subsided, have been severely and permanently damaged. Some in need of medical care haven't been able to get to the doctor. Several people have died; others were made homeless by the extreme weather events that have rocked this region.

On the Pine Ridge Reservation, people went without drinking water for weeks. They were only able to receive it, and food and supplies, via helicopter or horseback.

Farmers and ranchers from Nebraska, Iowa, Wisconsin, Minnesota, and South Dakota have also been hit hard by this disaster and its effects. Millions of acres have gone under water, killing livestock.

This winter and spring, I've witnessed firsthand that climate change is real. Higher global surface temperatures raise the possibility of more droughts and more intense storms.

You don't have to take my word for it—this deduction was drawn by the U.S. Geological Survey. But elders of Indigenous nations have been warning us about this for years.

In fact, a few years ago, Indigenous elders in Canada gathered to discuss climate change, its dire effects, and the desperate need to address it. As one who practices Oceti Sakowin (Great Sioux Nation) spirituality and observes my ancestral rites, rituals, and ceremonies,

I've been present many times as medicine men have warned us of the destruction of Western civilization and life on this planet if we do not unite and address environmental ruin and climate change. As the planet warms, even ceremonies like the Sundance, one of the seven sacred rites of the Lakota wherein participants dance, offer flesh from our bodies, and pray from sunrise to sunset for four days without food or water, become more dangerous for us to complete. Yet we know we must because it is who we are, and it's our way of maintaining universal harmony.

Our understanding—firsthand or otherwise—is backed by world-renowned climate scientists, who now persistently remind us what's at stake. A report by the United Nations Intergovernmental Panel on Climate Change, which was released last fall, warned that we must act to stop the planet from warming more than 1.5 degrees Celsius (2.7 degrees Fahrenheit) in the next twelve years or humanity will face historic heat waves, droughts, and floods, and hundreds of millions of people will be plunged into poverty.

Some point to evidence that shows it's already too late to stop global devastation, and that if we act immediately and aggressively, the best we can do is prepare ourselves to survive.

We appear to be headed for a 3-degree-Celsius increase. A study released in October concluded that the world's oceans had absorbed 93 percent of warming caused by humans since the 1970s, and global warming is far more advanced than previously thought. To make it clearer: A 2-degree-Celsius rise in temperature would flood out dozens of the world's coastal cities, cause massive heat waves, and leave half a billion people in a state of water scarcity; at a 3-degree increase, southern Europe would become overwhelmed by drought, and wildfires in the United States would increase sixfold; an increase of 4 degrees would see grain yields drop by half, quite likely leading to worldwide famine, war, and migration. A 5-or 6-degree rise in global temps might cause the extinction of most life on the Earth. According to the International Energy Agency, our remaining reliance on fossil fuels will guarantee a 6-degree increase in global temperature. Life could not survive it. There are already 150 to 200 plant, insect, bird, and mammal species going extinct every day.

Sounds apocalyptic, right? A little hard to digest? Look around you. We are already experiencing global disasters that are caused, or exacerbated by, climate change. And the recent disasters I described do not come as one isolated period of extreme examples. While our reservations were flooding, Mozambique was slammed by Cyclone Idai. Tens of thousands were displaced and more than 847 people were killed. Idai laid bare cities and created an eighty-mile-long inland sea.

Last week alone, wildfires sprang up on the Northern Cheyenne Reservation. It's not even fire season. And over the past few years, Montana has struggled with extensive wildfires. So has California; fifteen of its twenty largest fires have happened since 2000. Some scientists have said the increase is likely attributable to higher temperatures and drought; Southern California has seen a rise in temperature of about 3 degrees Fahrenheit over the past century. And an additional ten million acres of western forests have burned as a result of climate change.

This week, more than 118 million Americans are at risk from another wave of intense storms expected to produce powerful tornadoes and wreak havoc.

And since Hurricane Maria's landfall, in September 2017, about three thousand people in Puerto Rico have lost their lives as a result of the damage it caused. Many experts point to the 2017 hurricane season to show that climate change is creating more powerful, deadly storms. Young people all over Mother Earth realize the urgency of this crisis. In March alone, millions have demanded action on climate change in more than two thousand protests in 125 countries. These global climate protests are ongoing and support the urgency Indigenous cultures have pressed upon this issue.

Here in the U.S., we face great opposition to action on climate change from our own government. President Donald Trump just signed an executive order that seeks to take away states' rights to say no to fossil fuel projects, and he has routinely denied climate change while showing allegiance to oil and gas companies, removing our nation from the Paris Climate Agreement in August 2018.

Congressional leadership is attempting to address climate change through the Green New Deal, a largely youth-driven policy solution that would transfer the U.S. to 100 percent clean and renewable energy, and break the fossil fuel addiction that's putting all life on the planet in peril. The reality we face is daunting, but surrender is not an option. Supporting these new initiatives in the street and at the ballot booth is crucial to salvaging any livable future we want for ourselves and generations to follow. We cannot build upon old, outdated policies that serve only the 1 percent while ignoring the reality of global disaster and mass extinction.

This Earth Day, let us vow to do more than share the usual memes and celebratory greetings on social media. If we truly love this planet and all the life upon it, and want to see our children live, we must act. Now.

# The Fossil Fuel Industry Is Worsening the Global Plastics Crisis

## MAIA WIKLER

*December 21, 2018*

The global plastics crisis has mobilized action around the world, with governments implementing single-use-plastic bans and grassroots groups addressing plastic pollution in the ocean. But these efforts alone cannot address the sheer magnitude of plastics that are saturating water, food chains, and ecosystems worldwide. Only an end to the fossil-fueled plastics industry can.

The crisis is far greater than a consumer-behavior issue, like recycling: it is directly connected to the fossil fuel industry and to climate change, as 99 percent of plastics are derived from chemicals found in fossil fuels. Despite a recent United Nations climate report that says we only have twelve years to radically transform our entire economy to prevent the worst possible impacts of climate change, plastic production is set to ramp up, tripling the amount of plastic exports by 2030.

The future of the fossil fuel industry depends on plastics, and in the U.S., the recent rise in cheap shale gas from fracking is driving the plastics boom. Since 2010, over $180 billion has been invested into new plastics production plants that convert natural gas into ethylene, which is used to make many plastics. This means that the fossil fuel industry, responsible for pipelines that fuel climate change, is also responsible for the plastics crisis. When water protectors and activists resist pipelines across North America, organize for divestment from fossil fuels, and campaign for renewable energy and urgent climate action, they are also actively addressing the plastics crisis at the root.

"Fossil fuels and plastics are not only made from the same materials, they are made by the same companies," Steven Fei, an attorney for the Center for International Environmental Law (CIEL) Climate and Energy Program, said in a recent CIEL report. Some of the big players in the fossil fuel industry driving plastic production are Dow-DuPont, ExxonMobil, Shell, Chevron, BP, and Sinopec.

Petrochemicals come from oil and natural gas feedstocks, which create a wide array of products, mainly plastics. According to a report in the *Oil & Gas Journal*, plastics and other petrochemical products will drive global oil demand by 2050. A recent International Energy Agency (IEA) report, *The Future of Petrochemicals*, shines a light on the petrochemical industry as one of the blind spots in the global energy debate, and also says petrochemicals will be the key driver for oil-demand growth. It is expected to account for more than a third of global oil demand by 2030, and nearly half of oil-demand growth by 2050, according to the *Oil & Gas Journal*. Oil demand for transportation is expected to slow by 2050, due to the rise of electric vehicles and more energy-friendly engines, but that will be offset by the rising demand for petrochemicals and plastic production.

From the extraction of fossil fuels for plastics production to waste and degradation, the life cycle of plastics threatens the health of the environment and people. Fracking, a process that involves the high-pressure injection of water and chemicals into shale rock, cracks open shale rock to release natural gas. Plastics production is fossil-fuel intensive and carbon-heavy; the extraction and refining of fossil fuels by fracking contributes to global warming through greenhouse gases emitted by leaks. Not only does the production of plastics release methane, a greenhouse gas, but also plastics continue to release greenhouse gases as they degrade in the environment, which directly contributes to climate change. In turn, this affects sea level rise and and ocean ecosystems, while increasing severe weather catastrophes such as wildfires, drought, and flooding. Once plastics are produced, the pollution impacts are staggering.

Today, we produce about 300 million metric tons of plastic waste every year. That's nearly equivalent to the weight of the entire human population. Most plastics pollute the environment for significant periods of time, quite often breaking down into smaller plastic particles that can be swallowed by animals and fish and end up in our food and water. If current trends continue, our oceans could contain more plastic than fish by 2050. The life cycle of plastics shows that the fight against plastics pollution needs a more holistic approach. If we are going to effectively tackle climate change and plastics pollution, we need to stop plastic production at the source, which means transitioning away from fossil fuels.

A recent report in the *Oil & Gas Journal* said that regardless of whether aggressive action is taken to pursue renewable energy in the face of climate change, the plastics sector almost guarantees growth for oil and natural gas. The industry knows that plastics pollution is vast: it reports that the amount of plastics in the Pacific Ocean covers an area three times the size of France, while microplastics accumulate below the surface and enter food chains. Because of growing awareness of climate change and plastics pollution, corporations cite their support for recycling, but some would argue it's a marketing strategy to help maintain the longevity of the fossil fuel industry.

"The movement to address single-use plastics is a necessary part of the strategy to end plastics pollution, but it isn't enough unless there is widespread recognition that the plastics industry continues to expand," Carroll Muffett, CIEL president and CEO, tells *Teen Vogue*. "The recycle triangle is effective in convincing consumers that they aren't throwing something away, that it will be reused. But the truth is, less than 10 percent of plastics are effectively recycled."

While corporations market recycling and waste management, they are also lobbying against plastic regulations. The American Chemistry Council (ACC) represents companies like ExxonMobil and Dow and is reportedly responsible for lobbying efforts that successfully convinced the California Department of Education to edit environmental-curriculum textbooks to include positive statements about plastic bags, and also lobbied against a plastic-bag ban in the state.

Some of the companies fueling the plastics crisis are the same as those accused of climate change denial and muddling conversations on the subject. Exxon has spent millions to "confuse the public on global warming science," according to a report by the Union of

Concerned Scientists, and to prevent the U.S. from engaging in early implementations of climate agreements, like the Kyoto Protocol. The fossil fuel industry is now touting its support of recycling through marketing campaigns in order to keep the plastics sector booming.

There are current plans for a $10 billion plastic production facility to be built in the small town of Portland, Texas, by ExxonMobil and Saudi Basic Industries Corporation (SABIC), and many in the community are resisting fiercely, in defense of their health.

"We took Exxon by surprise. They thought they were going to come in and convince everyone [that the plant was a good idea]," Errol Summerlin, an organizer with Portland Citizens United, tells *Teen Vogue*. He says the proposed site would be located across the street from homes and within one mile of local schools. (*Teen Vogue* has reached out to Exxon for comment but has not yet received a response as of the time of printing.) "We're up against billions of dollars, Exxon and the Saudi royal family," Summerlin adds. "We have elected officials saying it's a wonderful thing for our local area, and we 100 percent disagree."

Summerlin claims that the amount of freshwater the plastics production plant requires, which he says is 7.3 billion gallons a year, is more than the 7.1 billion gallons all of the region's residents use. He says that local counties have been on water restrictions because of ongoing droughts and says further that the government is proposing seawater-desalination facilities to supply more water for industry.

Many industrial sites are being built in the backyards of marginalized low-income communities that may face health impacts from air and water pollution. These frontline communities have to pay the price of plastics expansion with the quality of their lives.

Priscilla Villa, a Hispanic third-generation Texan and organizer for Earthworks, a nonprofit organization dedicated to protecting communities and the environment from the adverse impacts of mineral and energy development, works on the front lines of Karnes County in Texas, one of the topmost oil-producing counties in the state. "People have been reporting health issues every day since the fracking boom. They have to live with nosebleeds, chronic headaches, trouble breathing, skin reactions, burning eyes, sinus problems," she tells *Teen Vogue*.

Adelita Cantu, PhD, RN, associate professor at UT Health San Antonio, School of Nursing, says these health problems may be related to fracking and the oil industry:

> As a public health nurse, I was concerned about how the environment impacts our health. What's happening in Karnes County and similar areas with oil and gas drilling and fracking, there are big health concerns. In terms of reducing quality of life and life expectancy and overall what it does for people's well-being. There are huge health consequences from fracking and oil and gas drilling. We are working with individuals in the community to be citizen scientists, giving them air quality monitors so they can document what is happening in their environment.

Cantu continues,

> I've been a public health nurse for over forty years and the environment does impact health and plays a role in quality of life. Knowing that connection is vital to the well-being of our communities. It's imperative for us all to think about our environments when thinking about our health. I've interviewed people in Karnes County who certainly have nose bleeds, skin rashes, and

respiratory issues—these are the main issues that come about from fracking and oil and gas sites.

Villa describes one plastics facility as being the size of a football field, illuminated twenty-four hours a day, seven days a week, and just four hundred feet away from a family's back door. Villa adds that these sites are constant sources of noise and light pollution and that people have reported suffering from insomnia. Dr. Cantu tells *Teen Vogue*, "One man I've talked to has an oil well in his backyard and suffers from respiratory issues and insomnia, related to the fact that lights are on all the time at the drill site."

"Community members are fed up with living near these sites and wish they could move, but how could they ever sell their house? No one would buy a house there. It's inescapable," Villa says. Community members feel that the government is prioritizing industry over the health of local communities.

The plastics boom means more fracking and greenhouse gas emissions that contribute to climate change while continuing to pollute oceans, food chains, and water supplies. It will perpetuate a fossil fuel economy that undermines efforts to address the climate and plastics crisis, and impacts frontline communities and the wider public at every stage of plastics' toxic life cycle. With only twelve years to prevent the worst possible impacts of climate change, the world cannot afford an expanding plastics industry in the time of crisis.

# Recycling Isn't Going to Stop Plastic from Destroying the Earth

ISABELLA GOMEZ SARMIENTO

*December 20, 2018*

You're probably familiar with the recycling symbol: the triangular chasing arrows that appear everywhere from your to-go cup to the plastic containers used to pack up leftovers at a restaurant. But although the sign might make it seem like almost everything we use has the potential to be recycled, the truth is, a lot of the items we place in our recycling bin may not end up where we expect them to.

Not only is it difficult to recycle each type of plastic in every community, but consumers also choose not to. The rest ends up accumulating in oceans or landfills, which disrupts natural ecosystems and degrades the planet. It's a common misconception that recycling can make up for how much plastic we consume—and even if everyone chose to recycle all the plastics they consume, the process is not as efficient as we may think.

"Recycling can honestly be really confusing if you don't put in the time and effort to research about it. I never knew that I had to wash all the food out of plastic containers to recycle it," Mishka Banuri, an eighteen-year-old high school student and environmental organizer from Utah, tells *Teen Vogue*. "For years, the plastic I recycled just ended up in landfill because there was food on it. My frustration with recycling is that it isn't straightforward, and that can make it inaccessible to many people who don't have the time to know about their local recycling regulations."

Ideally, recycling is the circular process in which materials that have already been used are broken down and repurposed into new products so as to cut back on the energy and resources required to produce virgin, or all-new, materials. Since there are different kinds of waste—such as plastic, paper, glass, and aluminum—there are different factors that determine whether a product will ultimately be recycled.

"The problem is that recycling is a business, so it's economics that are going to influence whether something is recycled or not much more than the technical ability for that material to be recycled," Stiv Wilson, a plastics pollution expert and director of campaigns at the Story of Stuff Project, an organization focused on environmental and social change, tells *Teen Vogue*. Although the current state of recycling partially depends on having a market for recycled waste products, that wasn't always a driving force behind the process.

The practice of reusing scrap objects can be traced back to ancient civilizations and was common throughout colonial times, when certain materials could be hard to come by and families tried to get maximum use out of household items, according to historian Susan Strasser's book, *Waste and Want: A Social History of Trash*. During the nineteenth century, peddlers who traveled from place to place selling goods bought discarded items that they could salvage and resell, establishing the first markets for used materials.

Later, as part of the World War II effort at home, the U.S. government also incentivized Americans to gather scrap metal, rubber, and other materials to be turned into weapons and other items for the war (although critics now claim the scrap drives served as propaganda to fuel patriotism more than environmental conservation). After the war, the economic boom of the late 1940s and 1950s created a major shift to a consumer-based "throwaway" culture that saw the introduction of household, disposable plastic products—eventually leading to the development of waste and recycling collection to try to keep up with the resulting pollution.

Today, most people are familiar with curbside recycling, in which local governments or private companies gather a neighborhood's recyclable products, take them to facilities, sort them, and reintroduce them into the system. But despite the establishment of these recycling efforts decades ago, only 9 percent of all the plastic produced since the 1950s has been recycled, and the United States recycles and composts just 34.7 percent. What doesn't get recycled can end up in dumpsites, landfills, or incinerators, which use heat to burn waste and release carbon dioxide into the environment. Although incinerators are believed to follow a "waste-to-energy" model by generating electricity, environmentalist organizations argue that they create a higher demand for waste and pose health risks due to high emissions of pollutants.

The supply of plastic has skyrocketed in the last sixty years, too. The global production of newly manufactured virgin plastics jumped from 2 million tons in 1950 to 381 million tons in 2015, and companies like ExxonMobil Chemical and Shell Chemicals are investing in new facilities that will reportedly increase plastic production by 40 percent in the next ten years. These plastics, which can take around 450 years to decompose and which release greenhouse gases if exposed to sunlight, are of immediate concern to environmental advocates.

Although the blame for pollution is often placed on consumers of single-use plastics, experts like Wilson and Martin Bourque, executive director of the Ecology Center in Berkeley, California, highlight the role that corporations play in environmental degradation. In hundreds of cleanups and brand audits led by the Break Free From Plastic movement, Coca-Cola, PepsiCo, and Nestlé were found to be the top plastic polluters worldwide.

"A lot of the problem with what we call contamination in recycling is people hoping or thinking that something is recyclable or being told by the manufacturer or the brand owner, the company that they're buying from, that it's recyclable," Bourque tells *Teen Vogue*, "when in fact there isn't any good way or any economic way to recycle the product or the packaging that's being sold."

By using the recognizable recycling symbol, he says, many companies push the idea that their products are okay to consume because they're technically recyclable, even though the infrastructure or market to recycle them may not exist. Multi-layer packaging, like in disposable coffee cups or health-bar wrappers, is especially problematic because it's the combination of different materials that would need to be separated—a complicated and time-consuming process—in order for the products to get recycled.

"Ever since I was little, I was always told that recycling is the best thing I could do to save the environment. But in reality, it's just another way neoliberalism places burdens on the individual rather than the systems that are responsible," Mishka says. "In short, recycling is another way to live a more sustainable lifestyle with a smaller eco-footprint, but individuals recycling isn't going to stop the problem."

Because not all local facilities have the capacity to recycle certain types of materials, much of U.S. waste gets shipped to other countries. According to a report by the Global Alliance for Incinerator Alternatives (GAIA), the U.S. is the largest national exporter of plastic waste worldwide. Until recently, China was the top importer, but in January 2018, the country implemented a ban on plastic scrap imports to de-escalate pollution. The result has been waste buildup in many Western countries as well as the rerouting of their pollution to Southeast Asian countries like Indonesia, Thailand, and Malaysia. Beyond creating environmental disasters, plastics prove to be a political issue, too.

"Part of the problem is that in some instances, waste that cannot be recycled, they're actually sent to these countries in the guise of recycling. But when they arrive here, they are either not recyclable or the infrastructure is not existing for it to be recycled," Froilan Grate, the executive director of GAIA Philippines, tells *Teen Vogue*. "So in a way, what happens is it ends up being dumped in these communities."

Informal waste pickers then try to sort through the mixed waste for valuable or recyclable materials, he says, becoming exposed to toxins and other harmful substances along the way. Whatever can't be recycled is left as pollution, oftentimes contaminating the bodies of water that the island communities rely on for their food. Even though Asian countries are often rated as the top polluters in the world, says Grate, these communities are trying to reduce their waste—they're just burdened with incoming plastics from the U.S. and Europe. According to Wilson, many corporations purposely "market the convenience-industrial lifestyle" to Southeast Asian countries, knowing they lack the environmental controls and facilities to handle the waste and then blaming them for littering.

Wilson, Bourque, and Grate all agree that there is simply too much plastic for it to be recycled away. Instead, they say, it's necessary for us to refocus on the first two of the three R's (reduce, reuse, and recycle): mainly reducing but also reusing. Scaling down on takeout that produces disposable waste, shopping in bulk with reusable containers at the supermarket, and carrying reusable water bottles to avoid plastic ones are all good steps we can take.

However, Wilson says, "it's a terrible place to stop."

The key is to call for collective action to hold local recycling facilities, government, and large corporations accountable for the production of single-use plastics and the way our waste is handled in the longrun. Joining the Break Free From Plastic movement, supporting organizations that fight back against plastic pollution, and contacting your local and federal government about environmental legislation are all important ways to stay engaged in the search for solutions. That's exactly what Mishka and her peers did in their state, lobbying lawmakers for two years until Governor Gary Herbert signed a resolution acknowledging the effects of climate change in Utah.

They went on to host a youth summit and launch their own organization, Utah Youth Environmental Solutions (UYES), in order to continue pushing for a healthier planet. "My biggest piece of advice would be to join local youth organizations who are involved in grassroots organizing. Organizations like UYES, Sunrise, Power Shift, or iMatter are all national organizations that can work on local issues in your community," she says. "Social media is also a huge tool that I've used to find opportunities to network and get engaged with youth who are doing similar work."

# I Went to the
# Great Pacific Garbage Patch.
# This Is What I Saw.

**ALLI MALONEY**

*December 22, 2018*

The Great Pacific Garbage Patch (GPGP), a site of marine debris considered to be twice the size of Texas, is perhaps the foremost expression of the impact of plastic waste on our world and the role of humans in environmental degradation.

It has been popularized through media coverage as the world turns its focus to plastic pollution, but misrepresented by mis-attributed photos that claim to show matted, flat surface debris in the middle of the Pacific Ocean. It is incorrectly believed to be visible from space and described as the "world's biggest landfill"; a so-called trash vortex where plastic is "piling up."

But it's just one manifestation of the many ways man-made environmental destruction has taken phenomenal hold of our natural world. Its alleged dramatic aesthetics fail to fully address the impact of the waste and the root of the global plastics problem. So, to understand its mythology and get to the bottom of what the GPGP really means for the planet, I went to see it for myself.

It takes over one thousand miles from shore to get there, departing from the West Coast and straight into the Pacific. Land fades from sight and the world around the ship becomes only water and sky. I set out this past September from Ensenada, Mexico, with a photographer to bear witness as a guest of Greenpeace, the decades-old nongovernmental environmental organization whose oceans campaign team conducted research from aboard their icebreaker, the *Arctic Sunrise*. The twenty-one-day-long expedition at sea shed light on and debunked prevalent ideas—mainly that the ocean, in any part, can be "cleaned up" from the mess humans have made.

We traveled directly toward the gyre, stopping only once for the engineers to make midnight repairs to the ship. Upon arrival, which took days, I expected to see trash everywhere, piled high as I heard it would be. What I saw was different and certainly no island. As Greenpeacers described to me, and as I witnessed, the GPGP is more of a "soupy mixture," with its most buoyant pieces of large, tough plastic joined by fishing debris at the very top of the water's surface and countless microplastics immediately—indefinitely—below. There was no oversized heap like I was expecting. There was no matted debris. Just vast sea, a few seabirds, and a touch of marine life amid a noticeably high concentration of waste.

It's home to a severe problem and is a visible manifestation of "throwaway" culture, wherein much of our economy and daily lives rely on plastics, most of which are disposed of after one use.

The GPGP was discovered in 1997 by marine researcher Charles Moore and named by oceanographer Curtis Ebbesmeyer. It became known as "Trash Isles," thanks to a pair of advertisers who appealed to the United Nations to have the area become the world's 196th country on World Oceans Day in 2017. The campaign was marketed well, and public understanding of the GPGP was generally founded on the notion that an "island" of trash had been discovered.

That misconception installed the impression that the impact of plastic pollution will be visible to the eye. The area is in the North Pacific Ocean between California and Hawaii, twelve hundred nautical miles offshore, where very few have ventured to bear witness, so widespread misunderstanding persists.

It is in the largest and perhaps most well known of the world's five ocean gyres, or systems of circulating ocean currents. It is one of three major "garbage patches" found within these gyres where, over time, plastic debris has coalesced. The mass of trash hits its peak in the center of the GPGP's most concentrated area, which fluctuates with conditions. Ships can enter easily, but even in its outermost zones, floating plastic debris appears with great frequency.

With the ship slowed down from its usual nine knots, the Greenpeace team spent an hour each day with a special trawl net lowered into the water. We'd sift the plastic pieces that were caught and pick them out of a tray and onto a gridded sheet to be counted and examined one by one, using tweezers. (The process, which feels endless, was oddly satisfying.) Members of the oceans campaign team then documented and packaged up the day's tiny finds to be sent to partner scientists to study and ideally trace back to a particular product or brand. On our first day of the sixty-minute practice, 1,119 pieces were captured and cataloged.

To better visualize what that looks like below the surface, they also needed to send divers. Tavish Campbell was one of two aboard the *Sunrise* and tasked with filming underwater. Before the trip, he'd seen "images in the media which made [the GPGP] look like a massive island you could walk on," he tells *Teen Vogue*. "I had prepared myself to see vast tidelines of plastic floating on the surface, complete with entangled sea creatures, but what we actually found was a far different story."

Instead, he encountered a seascape that he describes as "sinister": a vast expanse of pristine-looking ocean found to be "awash in trillions of micro-fragments of plastic" below the water.

"Every time I ducked under the surface into the bottomless blue, I could see tiny pieces of plastic drifting around me, some smaller than sesame seeds and hardly identifiable, but always present," Campbell says. "I have dove along shorelines thick with plastic garbage in the western Pacific and have witnessed the careless dumping of garbage closer to home in the eastern Pacific, but seeing the GPGP really linked it all together for me and offered a startling realization: there is no 'away' when something drifts away. It just heads toward the closest ocean gyre."

The majority of the plastic in the ocean eventually sinks. Smaller pieces float to the top, like those Campbell saw. This fragmented plastics problem is pervasive in ocean and drinking water around the world, Greenpeace senior oceans campaigner David Pinsky tells *Teen Vogue*. So much so, he says, that "the Environmental Protection Agency's samples [from its] water on-site, had microplastics in it."

Microplastics—which, as they sound, are miniature pieces of plastic less than five millimeters long—have been found in human feces, as we eat them in fish and most table salts. While systems of measurement have not yet been established to unify the world's research, in the GPGP, microplastics have been found to make up 94 percent of the pieces of plastic in the gyre.

Free-roaming man-made plastic matter can devastate the earth. It can lodge itself onto or into sea life not meant to carry or eat it, and can get stuck inside their bodies and cause choking. Microplastics are consumed by wildlife at high rates, with a recorded impact on at least eight hundred species, including half of the world's sea turtles and an estimated 60 percent of all seabird species, a figure predicted to reach 99 percent by 2050. Plastic ruins soil, leaching contaminants into the ground and waterways, and encourages pathogen growth, which can destroy reefs. When plastics large or small decompose in the sun, they release greenhouse gases that further advance climate disaster.

When we on the ship weren't trawling, we'd keep watch for plastic from the ship's side during the day or hit the sea in smaller boats to pull bigger pieces that were potentially branded or stamped and could lead to corporate accountability—a major part of Greenpeace's current mission, which asks the world to consider what "thrown away" actually looks like. Water samples were also taken in search of microfibers three to five times a day.

Microfibers are a major part of the plastics crisis, but only recently discussed. These microscopic particles, which shed from textiles and are not visible to the human eye, pollute a majority of the world's tap water and are commonly found in bottled water (in the U.S., 94 percent of tap water samples in one study included the fibers). They come from both natural materials (like cotton) and synthetic (like spandex) and are "smaller than a human cell," Pinsky says.

The impact of synthetic fibers on human health is still unknown but being investigated, though it's already clear that the chemicals that make plastic are endocrine-disrupting compounds, which can mess with human hormones, manipulate the functions of organs, and are said to even contribute the presence of ADHD in children.

While fragmented microplastics and minuscule microfibers are still being researched, we already know that the sheer volume of pollution they contribute to and represent is disrupting the planet. It's why many reject the notion that the ocean can be "cleaned up" by simply scooping up the plastic and carrying it back to shore. (A highly publicized, extremely expensive "cleanup" effort has even been set into motion by The Ocean Cleanup, a Dutch nonprofit, to little-reported success thus far.) There is simply too much plastic and most of it too small to capture.

Not all plastic in the water is micro; there is material you can see with your eyes, which gathers in the GPGP and can't be missed on beach shores around the world. We noted these from the side of the ship during most daylight hours. Talking at sunset one day with engagement coordinator Dan Cannon about his career with Green-peace, which started when the young organizer was a student, conversation was frequently interrupted to keep count—"another one," "there's two more"—of the plastic we'd speed past.

Life aboard a ship is equal parts exhilarating and exhausting. Living on the *Sunrise* gave me abs—as an icebreaker, it both pitches and rocks side to side, so much so that the crew calls it "the washing machine," and I was constantly holding on or gripping to stay steady. Each day, we'd get a 7:30 a.m. wake-up call in our bunk beds from Myriam or Robin, two millennial Americans who worked night watch while we slept. Chores were at eight, lunch at noon, and dinner at six, with all meals prepared by Daniel, a talented chef from Mexico City, with help from Amanda, a Hawaiian punk who runs a kayak shop in Seattle, or Pablo, a deckhand from Argentina.

The sea belongs to no country—it's an international rule—and the Greenpeace team embodied the notion that our environmental efforts should not either. Our radio operator, Rosy, hailed from Brazil, and Cat, the Italian medic, speaks six languages. The first, second, and third mates were from Finland, South Korea, and South Africa. Other crew and campaigners onboard represented Chile, Bulgaria, New Zealand, Canada, Belgium, Great Britain, and France.

In the GPGP's most concentrated zones, we'd venture out at least once a day in the smaller vessels that the *Sunrise* housed, lowered into the water by crane with a driver already inside (passengers would get into them through a door on the side of the ship, where we'd hold onto a rope ladder and jump in backward). I found myself with my hands in the sea, pulling out toilet brush handles, bleach bottles, laundry baskets, a suspension band usually found in hard hats. There was a disposable razor handle, hydrogen peroxide container, toolbox top, flower pot, water cooler lid, luggage wheel, buckets, a VHS cassette box with a fish inside of it, an unopened bottle of carbonated water, and a piece of Astroturf. White objects were the easiest to spot, but it came in all colors and shapes, fully intact, visibly torn.

The team recovered countless buoys, some as big as a beach ball, others small and compact. These were markers of the fishing industry's impact on the ocean, which weighs heavily. According to Ocean Cleanup, nearly 50 percent of the patch's total plastics tonnage is largely accounted for by fishing gear like plastic-lined nets that have drifted out to sea or been dumped into it, with much floating toward the area after Japan's 2011 tsunami. The *Sunrise*'s motorized crane lifted these "ghostnets" when we chose to stop and pull one from the water (an impressive, upsetting sight). Fish were to be pulled from the piles and thrown back. Crabs—of which there were varied species, riding on nearly every piece of plastic we pulled from the water—scuttled off, sealing their own fate.

It was hard to not feel the monumental weight of human failure as I spent day after day in the GPGP. Early on in the trip, U.S. actions director for Greenpeace Katie Flynn-Jambeck warned that "we might all cry" when we got there, and she was right. I did. I felt hopeless standing starboard-side on the *Sunrise*, counting my ninety-seventh piece of large plastic spotted in two hours on watch. Counting and organizing hundreds upon thousands of microplastics, tiny fragments that came bleached white, hot pink, and robin's egg blue alongside tiny bits of broken-down rope, I found myself thinking about the caps of pens, lids of yogurt, Barbie cars—plastic, everywhere, across the landscape of my life.

This realization was painfully reinforced when I was off the ship and hyper-aware of each product I saw for sale back home in New York City, where throwaway culture is key. While there have been proposals of banning plastic bags statewide and plastic straws in the city, the continued manufacturing and mass use of these will continue to pose economic and environmental issues for this other island of trash, where nonrecycled plastic is either buried or shipped to landfills in other states.

The solution, many experts now say, is to drastically slow down its production and consumption.

Plastics in the form of reusables like bottles and containers came into prominence among consumer goods after World War II as industries saw dollar signs and used chemicals to introduce new, cheap alternatives to other man-made products, which took skilled labor and natural materials to create. Today, we create three hundred million tons of plastic every year, half of which is for single use. We rely on it every day, in the clothes we wear, in our classrooms and offices, when eating prepackaged food and beverages, or shipping products by mail.

Long before it makes its way into a gyre, plastic causes problems. The creation of plastic products and its chemicals relies on fossil fuels, most of which are extracted from the earth in a ruinous process known as fracking. It is transformed through refinement for use, which contributes to global warming through leaks. It travels by way of pipelines, which are implanted into predominantly poor communities that are often exposed to pollutants. Plastic production itself is carbon-heavy and releases toxins into the environment. The facilities required for its creation are often built along waterways, which can flood in extreme weather and cause additional damage.

From start to never-end, plastic is dangerous. It demands land for resource extraction, production facilities, and waste storage, which has violent implications for Indigenous, marginalized, and impoverished communities.

As the problem intensifies, the most commonly proposed solutions are outdated. Recycling is important, but it is not enough to negate the impact of manufactured plastics on the environment: only 9 percent of all plastic ever created has been recycled. Packaging, which makes up for about a quarter of the total volume of all plastics used, is harder to recycle, as are colored plastics. (Biodegradables often need to be processed in facilities, too.) In the extremely frequent instance that a plastic material can't be recycled—an incineration process that requires energy and emits pollutants—it's dumped into a land-fill, where it will contaminate for more than one thousand years, or shipped from wealthy countries to others with less economic stability or political influence. People in these spaces pay the price of litter, pollution, and poisoning. Take, for example, Indonesia, the Philippines, Vietnam, and Sri Lanka: these are among the top countries considered "responsible" for marine debris, but are also some of the countries that receive much of the world's trash (and are then blamed for "mismanagement" of the overwhelming volume).

The politics of plastic are nuanced, and to deter the global crisis means to look beyond the recycling bin and toward "the corporations that got us into this mess," Pinsky says. "Companies have gotten [used to] a certain way of doing business and actually are pushing the cost off onto us, onto the commons, to our environment, into public health."

The plastics industry reportedly knew it was polluting the oceans back in the 1950s but only increased production, keeping consumers in the dark, Pinsky says. It has had great influence over regulations, has been on the receiving end of subsidies, and has long-standing, widespread lobbying power and deep government ties. Just like the plastics industry, the U.S. government appears to deny that the synthetics are related to health problems.

Up until this year, the U.S. sold its recyclable trash to China, exporting sixteen million tons in 2016. President Donald Trump failed to acknowledge that decades-long relationship (which also has economic ties) when blaming China for the ocean's plastic crisis while signing legislation in October, pledging a commitment to "clean [them] up." "As president, I will continue to do everything I can to stop other nations from making our oceans into their landfills," he said.

In the same year, the United States and Japan were the only two nations that refused to join the G7 Ocean Plastics Charter, a pledge to work toward 100 percent recyclable, reusable, and recoverable plastics and increase recycling by 50 percent by 2030. The Trump administration has shown no signs of slowing down the source of the crisis: the plastics industry. In fact, it has displayed quite a bit of support, from its move to reallow plastic bottles in national parks to environmental policy rollbacks that mark a committed partnership with the fossil fuel industry.

The industry is made up of everyday brands that are responsible for manufacturing billions of plastics and plastic packaging each year, largely single-use. There is little to no transparency as to exactly how much they create or distribute. An audit of plastic debris collected from six continents by the Break Free From Plastic movement, a group of over fourteen hundred organizations, found the world's biggest polluters to be a who's who of consumer culture. (Some of these brands spoke with *Teen Vogue* about their plans to combat the plastics problem in an additional story for this series, expressing "ambitious goals" to use reused plastic content or biodegradable products, but no plans to create less overall.) In the GPGP we pulled from the water still-branded, fully intact plastic vessels for items readily available at most pharmacies and convenience stores, products I'd repeatedly purchased and enjoyed prior to the trip.

Plastic seems unavoidable, especially when buying food at the grocery store, but Pinsky explains that the shop itself and the brands it stocks can avoid it and offer alternatives. Greenpeace has asked major supermarket chains to consider a full audit of all plastic products in their stores—a daunting, "almost impossible" task that gets them thinking about the overall issue. (Pinsky encourages those interested in combating plastics to hold their local chains accountable, too.)

Grocery stores have adapted before. Pinsky worked on Greenpeace's 2018 Carting Away the Oceans report, which has audited major chains for their seafood sustainability since 2008. The campaign has seen major changes happen over time, largely thanks to consumers and activists holding corporations responsible. All retailers in the first report received failing grades. By this year, twenty out of twenty-two passed, though at time of its publication, none of the profiled retailers had "major, comprehensive commitments to reduce and ultimately phase out their reliance on single-use plastics." Change could be on its way, however: just after the report was released in August, Kroger Co. (which operates multiple store banners such as Kroger, Ralphs, and Harris Teeter) promised to nix plastic bags in all of its stores by 2025 and plans to "divert 90 percent of [its] waste from the landfill" by 2020. Pinsky says that to show true commitment, a comprehensive plan to reduce single-use plastic must be released, too.

But as for the companies producing the products found on those store shelves, few attempts have been made to develop major innovative solutions, despite the well-documented problem. Pinsky says that if the grocery stores they've worked with are any indicator, it's in their best interest for leaders of every industry to start working on a fix to move away from fossil fuel–based plastics, and soon—their competitors may already be doing so because it's what this new generation of consumers demands. Meanwhile, their products, either plastic or packaged in plastic, are marketed to consumers as safe to use despite major and minor varied risks associated with its use.

Some company leaders are starting to look at recycled ocean-bound plastic as a source material because it's smart for their business's bottom line. HP and IKEA, for example, are both part of NextWave Plastics, a global business consortium focused on keeping plastics "in the economy and out of the ocean" that also includes Dell and General Motors. (IKEA has also promised to phase out all single-use plastics by 2020.) Beauty brands are starting to do the same. Food and fashion are both beginning to get creative to avoid contributing further to the epidemic as well.

Consumers have been putting the pressure on corporations to change production practices, including many young people rising to the challenge. "The brands that young people care about, those brands care about them and are trying to deliver products . . . and be hip and socially responsible as well, because they know that young people care about this," Pinsky says. "Younger generations can say, 'enough is enough.'"

Teen activists have organized their communities in demanding alternatives in schools and local businesses, and can call them out at any time on social media when they see branded plastic in a waterway or natural space, Pinsky says. In addition to bearing witness up close, organizations like Greenpeace are applying pressure in a myriad of ways, including a petition that asks major companies like Coca-Cola, Starbucks, and PepsiCo to "invest in alternatives and phase out single-use plastic."

It's time for protests and bans, to demand action from our law-makers—and it's on us, a world of people who've been conditioned to rely on plastics, to stand up in our own defense.

What you won't hear about the GPGP is that it's remarkably beautiful. That far out at sea—no distinct matted island in sight—the water is purple at its stillest, with neon ice-white-and-blue curls when crashing. It was refreshing to stand on the deck and imagine all the Pacific Ocean travelers before us; I found it romantic, as nature should be. But with each floating piece and microscopic sample, I was snapped out of my daydreams and faced again with the environmental crisis that modern humans have caused.

Plastic is unnatural and felt so there, as it does when seen in stream beds or forests. It's simply en masse, therefore dramatic, in this part of the world. Facing the world's crisis in its farthest-reaching corner forced me to remember our place and time in history. I could not walk across an "island," but I saw devastation in the Great Pacific Garbage Patch that imposed deep shame. Plastics are everywhere, ranging the gamut of size, more destructive and distressing than I'd ever imagined.

Without any immediate and drastic change to the way we produce and consume plastics, by 2050, production is expected to have quadrupled. This will exacerbate the ongoing climate crisis, running parallel alongside a projection that assumes average global warming since preindustrial levels could be about twice what it is now by then, too. Substantial transformation will take mass participation from individuals, governments, and industries. The damage and impact of plastic pollution is clear, but re-envisioning the future of consumption is an uncharted path. To activists like those in Greenpeace, it means seeing plastic as trash before it hits the Great Pacific Garbage Patch—while it's still on shelves, in every new beverage bottle or trinket we buy—and rejecting what's become normalized for something new: a plastic-free world.

# How Climate Change Is Impacting Animals in the Canadian Arctic

## ROXANNA PEARL BEEBE-CENTER

*December 1, 2017*

*In this reported piece, twelve-year-old writer Roxanna Pearl Beebe-Center explains what she learned about the impact of climate change on animals while visiting the Canadian Arctic.*

A distressed mother polar bear is stranded on a piece of ice in the middle of a great gray ocean in the Canadian Arctic. Eyes wide, she stares up at our ship, the *Ocean Endeavour*, a colossal, man-made block of steel that intrudes upon her wild, icy haven.

She leaps into the frigid sea, a cub at her heels, her eventual destination the shore about eight miles away, where she may spend a long time hungry. Seals are her favorite food, and she can only catch them from atop the ice. With the ice melting faster and forming later, she has less time to hunt.

The ice is moving in ways it should not. Scientists keep a sharp eye on the Arctic because it is our planet's air conditioning, cooling the globe. And if it malfunctions, as is already happening, people everywhere will notice.

Traveling with my family on a two-week trip to Greenland and the Canadian Arctic in July, I saw the effects of these changes firsthand.

It's where I met twenty-three-year-old Jennifer Kalibuk of Iqaluit, capital of Nunavut, the newest territory of Canada. She's an Indigenous Inuit woman, and climate change is affecting her food, traditions, and even her home, buckling porches and cracking ceilings. Permafrost—a layer of soil that is supposed to stay frozen—is melting. Pillars that are embedded in the earth to hold up Arctic buildings are sinking, causing large and costly damage. "It's impacting people's homes who don't have money to pay for the repairs," Jennifer tells *Teen Vogue*.

It's also getting harder to find food. During a recent winter, when Jennifer said the weather was "weird," her uncle nearly died while hunting on the ice at a time when it would ordinarily have been at its thickest. It had appeared that way that day, but sadly, it was actually weak, and he and his snowmobile sank. "He was in the hospital for about three months, so he is actually now, because of that experience, too afraid to go on the ice," Jennifer says. "That prevents him from hunting for his family, for providing for his family the traditional country food that we love."

Traditional practices have also undergone the impact of climate change. Before her grandfather died, Jennifer interviewed him about his favorite spot for hunting seals, which was only reachable when the water was frozen solid. "It doesn't get that cold anymore, so it doesn't actually freeze anymore," Jennifer says. "The traditional knowledge that has been passed down orally, telling stories of that particular place, it was gone with him. He was the last of his family to have known all the traditional knowledge of that specific place."

A week before we were in Cape Dorset, ice barricaded the Inuit community. Ships normally can get through the Arctic for only about six weeks in the summer. The only way out most of the year is by snowmobile or airplane, according to what locals told me.

In July, the roads are ditches of mud. Clouds of mosquitoes swarmed Brianna Rowe, twenty-six, a shipmate I followed as she interviewed Inuit children and teachers about their lives, school, family, and traditions.

Brianna, who works for Reach the World and is director of the climate-change education project Disappearing Ice, connects classrooms with travelers via the Internet, so kids can travel the world virtually. She was using her vacation, as a student member of the Explorers Club, to connect U.S. classrooms with Arctic classrooms so kids can learn about one another's cultures—including how the U.S. is one of the main countries affecting the climate much farther north.

"The decisions that people make in the U.S. have a global impact," Brianna tells *Teen Vogue*. "It is important for everybody to be aware of their decisions, and their impact on the environment has an effect on the entire world's environment."

Kristin Gates, a park ranger who trekked alone across the Brooks Range, knows this all too well. Part of the year, thirty-year-old Kristin patrols Denali National Park by dog sled, and she's making a short film about plastics pollution and garbage in the Arctic. "Climate change is something that affects us every day during the winter months," she tells *Teen Vogue*. Frozen rivers are Alaska's highways, and when they don't freeze solid, travel becomes dangerous and difficult.

Unfortunately, mushing (a form of travel that relies on dogs to pull a sled), which Kristin was hoping to learn about firsthand, is now almost nonexistent. Dog mushing as a practice mostly died out after Inuit communities were settled in villages by the Canadian government in the 1950s, so she heard stories about the old days but has not found anyone traveling by dog sled. Nowadays, travel mostly uses motors and gas, in four-wheelers, snowmobiles, and trucks. Kristin enlisted some of us from the ship and others from the community to clean up the beach. In about an hour we collected seventeen bags of trash, a potential threat to seabirds and mammals.

People's lives are changing from our warming world—but is the future of adorable little animals just as dim? For birds, that's not necessarily the case, according to George Sirk, an ornithologist and lecturer who was aboard our ship. "Evolution dictates that the fittest will survive. If you can't cut it, then you won't make it. But then there will be other birds that make it, they'll come up from down south and say hey there's lots of food up here," he tells *Teen Vogue*.

But, according to Sirk's calculations, sixty thousand bird species used to flutter and fly about our world, and now only about nine thousand do, he claims. As Sirk said, it's worth noting that birders have different definitions of what a species is. Melting could actually benefit some birds. When the ice melts, the breeding season will lengthen, and there will be more to eat. That adds to the appeal for birds from the south to come live in the Arctic, which is good for them but not for the native birds.

At the trip's end, everyone aboard our ship went their separate ways. My family and Brianna headed home to the U.S., George headed to Canada. Jennifer stayed aboard. Kristin left for the Arctic

Circle Trail through Greenland. And Arctic life remained, but with a future in peril.

Even though the Arctic is far away, you can help protect what's left. Ways you can help are outlined in the World Wildlife Fund article "Five Ways to Help the Arctic as the Planet Warms," and at the Ocean Conservancy site.

# I Traveled to the Arctic to Witness Climate Disaster Firsthand

### MAIA WIKLER

*July 17, 2019*

efore I traveled to the Arctic, I considered it to be a remote part of the world where scientists on expeditions gathered samples and photographed glaciers. My limited understanding mirrored media depictions that show empty lands or pristine wildlife, a narrative that emphasizes a vast, untouched wilderness without the people who've long stewarded the lands.

But it's more than just polar bear scenes in the Arctic National Wildlife Refuge, an immense ecosystem of 19.6 million protected acres where humans and animals alike have coexisted for generations. It's home to an ongoing human and environmental rights crisis that follows an unprecedented move by the Republican-controlled Congress, backed by the Trump administration, which in 2017 passed a tax bill that mandated the Coastal Plain of the Refuge be open to lease sales to the fossil fuel industry for oil and gas development to

59

help pay for massive corporate tax cuts. If it does occur on this sacred land—the only piece of Alaska's Arctic that has been protected for its ecological importance—devastating environmental and cultural impacts will follow.

I went there myself to report this story for *Teen Vogue*, to meet those fighting to save these lands and see the impacts of climate change firsthand. My journey lasted two weeks, traveling first by way of a ten-seater bush plane to the first Arctic Indigenous Climate Summit in Gwich'yaa Zhee, or Fort Yukon, Alaska. Hosted by Gwich'yaa Zhee, the Council of Athabascan Tribal Governments, and the Gwich'in Steering Committee, the summit provided an unprecedented opportunity for community members and Indigenous leadership to explain what is at stake in the Coastal Plain region of the Refuge.

Many locals, scientists, and advocates refer to the Coastal Plain as the biological heart of the Refuge for countless living species: in its wildness, an ideal nesting habitat is found in its wetlands and food supply, safe from industrial impacts. Millions of birds migrate to the Refuge from continents and ecosystems across the world, and a significant number of Southern Beaufort Sea polar bear dens are also on these lands.

"The Refuge is one of the last places left in the nation where you can experience hundreds and hundreds of miles without a road or trail. It's home to one of our last intact ecosystems, where grizzly bears, wolves, polar bears, migratory birds, and caribou live in relation with the land as they have for thousands of years," Emily Sullivan, an organizer with the Alaska Wilderness League, tells *Teen Vogue*.

## The Arctic Wildlife Refuge

The lands we know as the Arctic are sacred to Indigenous Gwich'in people. They call the Coastal Plain "Iizhik Gwats'an Gwandaii Goodlit" or "the sacred place where life begins," referring to the more than two hundred thousand–strong Porcupine Caribou herd that migrates north each year to the Coastal Plain to give birth.

Gwich'in spokesperson and director of the Gwich'in Steering Committee Bernadette Demientieff explained this relationship at the summit, saying, "The survival of the Gwich'in depends on the survival of this herd. For thousands of years, we migrated with the caribou—we settled along the migratory route so we could continue to thrive."

When the Refuge was designated officially protected public land in 1960, it became, in essence, owned by the American public. After decades of bipartisan efforts to protect the wilderness, the Trump administration mandated a minimum of two massive lease sales of the 1.6 million acres of Coastal Plain to the fossil fuel industry by 2021 to offset tax cuts for corporations and millionaires as part of the 2017 Tax Cuts and Jobs Act. The administration is pushing to have these sales later this year, and the initial draft of its environmental impact assessment didn't consider oil spills or climate change.

Republican senator Lisa Murkowski led this provision, and in doing so, also amended the stated purpose of the refuge itself to include oil drilling—a stark contrast to the original stated purposes, such as maintaining environmental health, conserving wildlife, and protecting wilderness values. The move comes following Murkowski's repeated acknowledgments of the effects of climate change on Alaska. Under her legislation, lands leased in the Arctic National

Wildlife Refuge would be managed as a petroleum reserve, which undermines adequate environmental protection.

Nobody really knows just how much oil is in the Coastal Plains area. Before drilling can happen, seismic tests assess which areas have oil. The impacts of these tests alone could be devastating, according to recent reports by Arctic scientists, as seismic exploration sends strong shock waves into the ground. Tests occur on frozen tundra so the vehicles won't sink, but winter is also birthing season for Southern Beaufort polar bear mothers, who seek refuge on the Coastal Plain—and the species population has already declined by 40 percent from loss of sea ice. Seismic exploration exacerbates erosion and also the thawing of permafrost, frozen ground across the Arctic, which the Refuge has experienced since seismic testing occurred there in the 1980s.

The Gwich'in organized the summit to connect allies and community members from villages across Alaska and address these threats. Elders, who traveled by plane from the Southwest, and Gwich'in members from Old Crow in Canada, who came by boat, joined hunters, chiefs, community members, youth, traditional and Western scientists, lawyers, and grassroots organizers, who came together to share their research and personal experiences with the rapidly changing climate.

The Arctic is known as ground zero of climate change because it is warming twice as fast as the rest of the world. On July 1, Juneau broke a 110-year-old heat record. Unusually hot and dry summers are fueling rampant wildfires too: as of the first week of July, 120 fires were burning in Alaska, prompting evacuation orders and air-quality alerts. Many of the log cabins in Gwich'yaa Zhee are tilted, which a local explained to me as a result of climate change melting the permafrost and shifting the foundation. "We've got rapid warming the last few years," he said. "I've never seen stuff like this."

For three days at the Summit, I camped along the river. We gathered outside under an open-air wooden octagon structure overlooking the Yukon River, surrounded by aspen, cottonwood, and bushes of wild roses. To travel such great distances to gather in a remote village requires intention, and there I felt a profound sense of unity, purpose, and community.

Speakers shared concerns about rapid glacier melt, increasing wildfires, earlier and condensed flowering seasons that lead to mismatched timelines for pollinators. Gwich'in leaders emphasized how drilling would impact their identity, food security, and livelihood. Hunters shared their stories of noticing changes in animal behavior. Elders offered prayers and encouraged healing in the community. Western scientists explained that the rapid rate of Alaska's increasing heat is also exposing coastal communities in the region to storms and erosion from vanishing sea ice: at least thirty-one villages are in imminent danger from coastal erosion, and three of those villages must relocate soon or cease to exist.

To come together in the Arctic was fitting. The world's livelihood depends on this region because it acts as a global climate regulator. When ice melts here, it drives extreme weather across North America, Europe, and Asia, with newly open water creating more extreme temperatures and moisture for storms. Research shows that disappearing sea ice could be affecting continental weather patterns. Melting permafrost releases methane, which also accelerates climate change.

"Our lands are now slumping, entire lakes draining, and even entire rivers reversing as the permafrost and glaciers that held our ancient lands together are now melting and eroding at accelerating rates," Chief Tizya-tramm, from Vuntut Gwitchin First Nation in Old Crow, Yukon, testified at a congressional subcommittee hearing in Washington, D.C., held in March with other chiefs, tribal members, and politicians.

Proponents of drilling in the refuge claim the Gwich'in will not be impacted because they don't physically travel to the Coastal Plain, a generations-long choice made so as to not disturb caribou birthing grounds that the community consider too sacred to visit. What the proponents fail to recognize in this logic is that by impacting the caribou, they impact the Gwich'in.

Hunting and the ability to live off the land is crucial to the community. It can become nearly impossible to have access to affordable food otherwise: grocery stores aren't often accessible or affordable, and gas often costs more than $7 a gallon. Community members and leaders emphasized the connection between protecting the Coastal Plain and protecting their identity as a matter of spiritual, cultural, and physical survival.

"I know what it's like to live in the village and go without a grocery store," twenty-two-year-old Julia Fisher-Salmon, Draanjik Gwich'in, told me. "The Inupiak people from the upper coast can't eat their meat anymore because it makes them sick, the oil rigs give off toxic particulate matter. . . . Oil and gas [are] forcing villagers who depend on this land to leave, to move. It's all connected."

"In the winter you can see black snow on people's rooftops from the oil field [being] so close," Siqiniq Maupin, an Iñupiaq mother and leader, told me. "We have been advised that they have found such high levels of toxins in the bowhead whale that we have a higher risk of getting cancer by eating our traditional food. The world has made it so that our traditional food is killing us."

Oil development could end the way of life that has sustained the Gwich'in and neighboring Indigenous communities for twenty thousand years. Many feel the adverse effects of opening the refuge to the fossil fuel industry would be akin to cultural genocide.

Following the Arctic Indigenous Climate Summit, my reporting continued as I joined Julia and group of young artists on a six-day trip into the Arctic National Wildlife Refuge, a trip organized by The North Face to bring awareness to the issue, reminiscent of their first-ever expedition to the region in 1972 to protest a proposed pipeline. We took three bush planes to reach our base camp near the Sadler-ochit Mountains, flying over the Brooks Range along the way.

Knowing that the region is untouched by industry, I was surprised to see countless dark lines intersecting on the tundra below. Daniel, our pilot, identified them as caribou migration trails, imprinted by caribou following their ancestors' paths for thousands of years.

Seeing their paths from the sky, the refuge looked like caribou country. This is a birthplace, where mothers feel safe and called to give life, and the trails of caribou mark the landscape rather than the man-made roads. Amid a global refugee crisis, this place shows us the beauty of migration, something that should be a part of every being's right to move in seeking refuge.

As we began to land, thousands of caribou ran below us on the flat tundra, the pulse of the plains. I noticed several turned in tightly twisting circles; Daniel explained the animals were disoriented by the noise of the bush plane. I could only imagine the disturbance and impacts seismic testing, trucks, drills, and oil fields would have.

We set up camp along the banks of the Hulahula River. Our group hiked for hours each day, navigating thick willows and feeling the buoyancy of permafrost and tundra under our feet. On our last morning, after days of exploration, we woke to wildfire smoke: the glacial-fed river was swollen from hot cloudless days, torrents of water gushed over the rocks and banks. We loaded a bush plane and took to the sky in a haze.

In a century, when this economic tax bill will be long gone, we will not be able to re-create the refuge. It relies on our collective imagination to see these lands as more than a short-term monetary gain for corporations and the government. As the first chief of Arctic Village, Galen Gilbert, said, "The Coastal Plain is not just a place on the map; it is a foundation of our entire life."

There's still hope that this drilling can be stopped, however. There is a bill currently in Congress, the Arctic Cultural and Coastal Plain Protection Act (HR 1146), that could overturn the legislation that authorized opening this region to development. The Gwich'in Steering Committee, environmentalists, scientists, and grassroots groups are calling for solidarity and action to help support the bill, including signing a petition and amplifying through public actions. According to lawyers and environmentalists at the Arctic Indigenous Climate Summit, lease sales are anticipated to begin as early as October, and the bill could see a vote in the House this month. If you want to join the movement to save the refuge, donate to the Gwich'in Steering Committee and support Indigenous leadership. Follow the Alaska Wilderness League for updates. Urge your U.S. representative to support HR 1146 so that the Refuge can continue to be a sanctuary for generations to come.

Like their ancestors, the Gwich'in are not giving up, and neither should we.

# Publicly Owned Utilities Could Help Fight the Climate Crisis

## GRETA MORAN

*October 25, 2019*

Earlier this month, Pacific Gas & Electric, the investor-owned utility company that supplies power to much of California, cut off electricity to over seven hundred thousand customers. The company argued that such a drastic measure—the largest planned power outage in the state's history—was necessary to prevent wildfires.

Yet, for some activists, this bleakly framed choice served as another reminder that investor-owned utility companies are not positioned to manage our energy futures, especially as climate change raises the stakes.

In recent years, activists around the country, including in New York City, Boston, Providence, Chicago, Boulder, and Washington, D.C., as well as Northern California and Maine, have been working to transition utilities to public ownership, which would make them accountable to the public instead of investors. In Northern California, the Let's Own PG&E campaign emerged shortly after the Camp Fire, the deadliest fire in California history, which was caused by PG&E power lines. In other parts of the country, activists are fighting against utility companies' direct contributions to the climate crisis and systemic inequalities.

The fight in New York City has grown particularly heated over the past few months. In July, in the middle of a deadly heat wave, Con Edison intentionally cut off power to some of New York City's poorest neighborhoods, sparking outrage. Last month, at a rally outside the National Grid office in downtown Brooklyn, activists from groups like Stop the Williams Pipeline Coalition and the New York City Democratic Socialists of America (NYC-DSA) spoke out against investor-owned utility corporations' continued investment in gas pipelines. In particular, they're angry about National Grid's push for the highly contested, twice-denied Williams Pipeline that would cut through New York Harbor and proposed rate hikes from the two major utility corporations.

"Both ConEd and National Grid are proposing to raise our rates to expand fossil fuel infrastructure," Lee Ziesche, an activist with Sane Energy Project, told the crowd. "What they are proposing completely fails the climate test."

The ultimate goal for some activists is to abandon investor-owned utility corporations altogether and build a more democratic system in their place. Last year, NYC-DSA launched the Public Power NYC campaign to make the energy grid publicly owned. Building on this idea, on October 19, a coalition of grassroots organizations launched Movement for a Green New Deal, a campaign that will demand utilities be publicly owned as a key part of this transition. Giving the public control over New York City's energy future, activists argue, could lay the groundwork for the just, rapid decarbonization of the energy sector.

"We could de-commodify clean energy and guarantee it to all New Yorkers as a human right, much in the same way we already guaranteed clean water through our public water utility," said Amber Ruther, an organizer with NYC-DSA, during a recent hearing before the New York State legislature.

Ruther argued in her testimony that investor-owned utilities are not positioned to tackle the climate crisis. "The incentive structure for private utilities was designed to encourage them to build as much infrastructure as possible," said Ruther. "But now that incentive structure is obsolete, and it's preventing us from achieving our climate goals."

To replace investor-owned utilities, the Public Power campaign is calling for a large-scale public utility. This could mean the expansion of the New York Power Authority, the largest state public utility in the U.S., or the municipalization of private utilities. Once established, a public utility would be responsible for the major work of transforming the grid, Aaron Eisenberg, an organizer with NYC-DSA, explained to *Teen Vogue* in an email.

Alongside the large-scale utility, the campaign envisions smaller, community-owned energy projects. As a model, Eisenberg points to Sunset Park Solar, a cooperative solar array, which will be owned and operated by the same people who purchase its power in central Brooklyn.

The idea of publicly owned utilities is quickly gaining traction. Mayor Bill de Blasio recently suggested socializing the city's utilities.

In late July, New York City council member Costa Constantinides partnered with the NYC-DSA to host a town hall in response to ConEd's proposed rate hikes. Over one hundred people, including assemblyperson Aravella Simotas and public advocate Jumaane Williams, traveled in the pouring rain to attend the town hall, where many demanded public utilities.

But no one from ConEd or the New York Public Service Commission, which oversees the state's utilities, showed up.

Jamie Tyberg, an organizer with New York Communities for Change, sees publicly owned power as an important step in transforming our relationship to resources and energy, which have for too long been regarded as infinite.

"We are not going to be able to escape this crisis without replacing the system," Tyberg, who has been volunteering with the campaign, told *Teen Vogue*.

In response to the claims made in this article, Michael Clendenin, ConEd's director of media relations, wrote in an email, "We can maintain high reliability and have a clean energy future. Con Edison is committed to helping New York State achieve its clean energy goals, and through our Clean Energy businesses we are the second largest solar producer in North America." A spokesperson from National Grid responded to *Teen Vogue*'s request for comment by providing links to their plan for "investing in the natural gas networks making them safer and more reliable, advancing a cleaner energy future." (The idea that more gas networks will advance clean energy has no basis in climate science.)

## A Widespread Movement, With Old Roots

Public utilities are hardly a new idea—over two thousand public utilities already exist in the U.S.—but the growing demand for a more just energy system has lent this ownership model new energy.

This summer, Vermont senator and 2020 presidential candidate Bernie Sanders released his own version of an ambitious Green New Deal, which outlines a plan for renewable, publicly owned energy.

Sanders's vision may look something like the energy system in the (admittedly less populous) state of Nebraska, which consists entirely of public power districts, publicly owned utilities, and energy cooperatives. The state abandoned investor-owned utilities in 1946 and remains the only state to operate fully on public power. Since then, Nebraska has been able to maintain some of the lowest rates of electricity in the country.

This future could also resemble Sanders's own town of Burlington, Vermont, which has had a municipally owned utility since 1905. In 2014, Burlington became the first city in the U.S. to move to 100 percent renewables, while maintaining the same rates since 2009.

The city recently laid out a roadmap for an even more ambitious goal: moving to net-zero energy by 2030, which the city defines as "sourcing at least as much renewable energy as it consumes in energy for electric, thermal, and ground transportation purposes."

Public utilities were also a component of the Green New Deal's predecessor: Franklin D. Roosevelt's New Deal. The 1936 Rural Electrification Act was passed to provide loans to isolated rural areas to start their own electric cooperatives. The goal was simple: to give farmers, left in the dark by profit-motivated utility companies, affordable electricity.

"At the time, investor-owned utilities were not really willing to go into rural America" because the sparsely populated areas did not promise good business, explained Johanna Bozuwa, who works at the Democracy Collaborative.

While today's campaigns have broader visions, the main advantages of a public utility remain. They are designed to answer to the people rather than a profit motive. Public utilities have lower average rates than investor-owned utilities, and any revenue made goes back into the community, rather than the pockets of CEOs and private shareholders. For instance, Austin Energy has used its revenue to support streetlights, public safety, libraries, and parks.

Investor-owned utilities don't have the same incentive. "As a top line, investor-owned utilities have a terrible track record," said David Pomerantz, who works at the Energy and Policy Institute, a watchdog organization that tracks energy utilities. He points to the fact that investor-owned utilities are often the top contributors to political campaigns and spend millions on lobbyists. This year, National Grid even asked its own customers to help lobby for fracked gas.

However, Pomerantz cautions that public utilities are not blameless. Like private utilities, public utilities have trade associations, like the American Public Power Association (APPA) and National Rural Electric Cooperative (NREC). Both groups have fought EPA standards for clean air, water, and carbon pollution, Pomerantz explained in an email.

Bozuwa, of the Democracy Collaborative, sees the current moment as an opportunity to learn from public utilities' mistakes and push this model in a new, bolder direction. "Public ownership is built within the context of the people who are organizing for it," said Bozuwa. This grants the system its power and, historically, its limitations.

## A New Era of Energy Democracy

Today's emerging campaigns are calling for a new system of "energy democracy" altogether and share an overarching goal of climate justice. What this entails takes many forms.

For instance, the #NationalizeGrid campaign in Rhode Island has been fighting, alongside its coalition partner, the George Wiley Center, for statewide public power.

"It's a pretty simple piece of legislation [requiring] a very minor tax increase across the state, [that would mean] low-income payers would only need to pay a certain percentage of their monthly income as their utilities payment," explained Corey Krajewski, an activist with the campaign.

Boston's Take Back the Grid campaign has been focused on ensuring workers and communities of color in Boston benefit from the shift to a publicly owned renewable grid.

Similarly, Northern California's Let's Own PG&E campaign has been thinking through ways to ensure that the PG&E workers are included in the new energy system. One idea they've been considering is a proposed ballot measure to protect these workers' pensions, explained Emily Algire, one of the organizers in the campaign.

In Maine, the local power supplier has been accused of erroneously charging customers. In response to disproportionately high rates, a group of over six hundred rate-payers is suing the Central Maine Power Company, while also backing a bill to replace the two main power providers in the state with a statewide public utility, introduced by Representative Seth Berry. For Berry, it's a major opportunity to address the climate crisis.

"Basically, we have to electrify our entire economy, and we have to do that really fast," Berry said. "It's like the foundation that we're building our home on for the future."

SECTION 2:
ACTIVISM

# Climate Activism and Organizing Is Changing the World and Offering Hope

## ALLEGRA KIRKLAND

*May 2020*

This is what democracy looks like: thousands of young people—from rural farm towns and dense urban centers, frontline communities and Ivy League institutions—coming together to fight for climate justice.

*Teen Vogue* has documented the rise of the youth climate movement over the past few years, as it has grown from smaller, campus- and community-based efforts to a social media–fueled global powerhouse. We've followed the lead of the young activists elevating this issue and come at this story from the many different angles that now represent our core areas of coverage: social justice, immigration, youth activism, and inequality.

Today's youth climate movement harkens back to the spirit of the original Earth Day in 1970. Contemporary activists, like those who once helped secure the creation of the Environmental Protection Agency and the passage of the Clean Water Act, approach our environmental crises from an intersectional lens. They want to stop sea level rise and mitigate blazing wildfires, but they also want justice for those who have borne the brunt of environmental disaster for decades. They want to lift up the voices of women, people of color, and Indigenous communities. They want to create a climate justice movement that is forward-looking and truly representative.

So we've tried to tell the stories of who these activists are and why the important work they're doing matters. We've run stories profiling individual organizers and representatives of some of the major groups, like Extinction Rebellion and Zero Hour. We want to know what their personal backgrounds look like and what's driven them to act. We've listened in on their conversations with friends, and they've told us what they're thinking when they occupy lawmakers' offices or spend long days protesting on hot city streets.

For one of my favorite climate pieces we've ever run, Sarah Emily Baum spent weeks with the organizers of the 2019 New York City Youth Climate Strike. Sarah documented their planning meetings in rec room basements and watched them design signs and map out programming for the main event. Their work culminated in a massive global march that saw over four million people participate in what was estimated to be the largest climate protest in world history. *Teen Vogue* staff left the office early on the sunny September Friday afternoon to join the floods of people crowding the streets of Manhattan's Battery Park to watch Greta Thunberg take the stage.

And we profiled Greta herself—the enigmatic, eloquent, righteously angry Swedish youth activist who has become one of the most prominent faces of the global climate movement. Greta's School Strikes for Climate inspired countless young people to give up their childhoods and their schooling to take a weekly stand against the most pressing issue facing humanity. And she helped make room for those who felt hopeless in the face of a frightening future, and apart from their peers. As she told *Teen Vogue*, "People who are different are so necessary, because they contribute so much. Therefore, we need to really look after the people who are different and who may not be heard. We need to listen to those and to look after each other."

Beyond providing moral leadership, youth activists have been taking concrete steps towards change. They have fought for divestment and helped eliminate the use of single-use plastics on their school campuses. They've organized for the passage of a Green New Deal. And some even sued the federal government for enabling the destruction of the planet with their subsidies of the fossil fuel industry and other major polluters.

We know some degree of climate disaster is already baked in. The waters will rise, the wildfires will burn, the plains will flood. But all is not lost. Crucial work is still being done; hearts and laws are still being changed. And we will owe the future we inherit to the fearless, groundbreaking activism of the youth organizers who fought for us to have one.

# "We Should Care"

## *Young Climate Activists Share Why They're on Strike*

### SARAH EMILY BAUM AND LUCY DIAVOLO

*September 20, 2019*

Hundreds of thousands of people across the world went on strike on Friday, September 20, 2019. Upset by inaction by world leaders and eager to take action on climate change before it's too late, the Youth Climate Strike participants took to the streets to make their voices heard. *Teen Vogue* caught up with attendees at the New York City rally to learn why they're striking.

*Genevieve Rodgers*, 19, Columbia University class of 2022, from Sewanee, Tennessee

"It is beyond time for something to be done. This is our future. If we don't do something about it, no one will."

*Grace Goldstein*, 17, Stuyvesant High School senior, from Manhattan

"I'm striking for frontline communities and I'm striking for every generation that will follow ours. . . . We should care about ourselves, we should care about each other, we should care about the fact that climate change is affecting real people and communities right now as we speak. And if it doesn't affect you now, it affects you next."

*Shirel Salinas*, 16, from Norwalk, Connecticut

"A bunch of kids from our school decided to come here today because we were tired of not being able to have anyone speak up for us. We decided to come and have the school environmental club and Democratic club get together; we're all here now."

*Sophia Murphy*, 18, Binghamton University student of environmental studies policy and law, and biology, from Brooklyn

"I'm a freshman, but I'm taking time off to organize climate activism. I'm striking today so I can have a future and so I can graduate law school and have kids—so I can live to see 85. And so people who are in the position that I am in can have safe, healthy futures and time to live and not be terrified that they're going to die in the next ten years."

*Sharon Damian*, 14, from Queens

"We have so many problems, and we can't just make this one when it's so easy to control by recycling and making new solutions."

*Nowshin Quader*, 16, John Dewey High School, from Brooklyn

"The world is ending and we have to stop it."

*Teen Vogue:* What do you say to those who say striking is pointless or that this doesn't matter?

"We live in this world and it does matter. When we will be dying, you'll see it will matter," Nowshin adds.

*Vincent,* 15, from Manhattan, and *Chloé,* 15, from Brooklyn

*Chloé:* "I came out to strike today because climate change is gonna really affect our earth, and it's probably gonna shorten our lives because it'll shorten the planet's life, and when that happens we'll have no place to live. I also think we live in America and people come here to live the American dream, but how are we going to live an American dream, or a worldly dream if we have no world to live on?"

*Vincent:* "I'm here because a lot of bad stuff is going to happen to our earth if we don't do something about it—all sorts of natural disasters. Just look at Cuba and Florida. Why stay at school when I can spread the word with my homies?"

# Nine Teen Climate Activists Fighting for the Future of the Planet

*July 24, 2019*

From a dedicated episode on the HBO drama *Big Little Lies* to the policy platforms of 2020 Democratic candidates, the inescapable reality of the climate crisis is at the forefront of everyone's mind—and for good reason. Since 1880, the global temperature has risen by 1.9 degrees Fahrenheit, with a majority of this warming happening in the past thirty-five years.

While 1.9 degrees may seem marginal, this increase has led to major changes in climate and negatively impacts the environment. Scientists have reported a nearly 13 percent decrease per decade in the Arctic sea ice minimum, an increase in stronger, more destructive storms, and an increase in extinction and animal endangerment, all as a result of climate change.

As politicians begin to discuss life-changing legislation like the Green New Deal, another group of environmental activists have begun the fight for immediate change. Taking cues from their predecessors, Generation Z has taken on the enormous task of saving the planet from future destruction—and ensuring they have a future to look forward to.

Get to know some of the Gen Z environmental activists hoping to change the world, one step at a time.

## 1. Greta Thunberg, 16

A Nobel Peace Prize nominee, Greta Thunberg is the Swedish teen who's quickly made her way to the forefront of the climate-justice movement. In 2018, Thunberg spoke about the importance of climate activism in front of a group of world leaders at the United Nations.

"Since our leaders are behaving like children, we will have to take the responsibility they should have taken long ago," Greta said at the summit. "We have to understand what the older generation has dealt to us, what mess they have created that we have to clean up and live with. We have to make our voices heard."

Thunberg has since led a number of school strikes globally and has spoken in support of the Extinction Rebellion demonstrations. The teen recently announced that she would speak at the U.N. Climate Summit in New York in September and the COP25 in Santiago, Chile, but not in person. Thunberg doesn't fly at all due to the "enormous climate impact of aviation."

## 2. Katie Eder, 19

The executive director of Future Coalition, the largest network of youth-led organizations and youth organizers across the country,

Katie Eder organized two climate strikes and aided the formation of the #AllEyesOnJuliana campaign by age 19.

Eder told *Teen Vogue* her quest for environmental justice began in sixth grade, after she read Al Gore's *An Inconvenient Truth*.

"I was so confused, at that age, how a problem like climate change could be so large and yet it seemed like nothing was being done to address the issue," she said. "I thought that the only way this could be happening was because no one knew about it, and so for much of middle school I dedicated myself to educating people about climate change."

Eder notes that throughout history, young people have been the catalyst for change.

"This is the revolution that's going to save our planet," she said. "Our generation is not going to sit around as our futures are destroyed around us."

### 3. Jamie Margolin, 17

"Life as we know it is coming to an end thanks to climate change and rapid environmental destruction," Jamie Margolin told *Teen Vogue*. "As a young person, I am always asked and expected to plan for my future. How am I supposed to plan and care about my future when my leaders aren't doing the same, and instead leaving my generation and all future generations with a planet that is inhospitable and impossible to sustain civilization?"

The cofounder and co-executive director of Zero Hour, a movement dedicated to giving a voice to Generation Z on climate change, Margolin believes the key to all justice is climate justice, telling *Teen Vogue* that correctly solving climate change means "dismantling all the systems of oppression that caused it in the first place." The motivation behind her movement is to promote a brighter future, not emphasize a "giant existential crisis."

Since founding her movement in 2017, Zero Hour has organized several actions, including lobby days, protests, and becoming a full-fledged organization.

"We're not a movement that happened overnight," Margolin said. And it's not one that's disappearing any time soon.

## 4. Nadia Nazar, 17

Cofounder, co-executive, and art director of Zero Hour, Nadia Nazar joined Margolin in bringing this environmental movement to life. Together, the girls helped lead a three-day event that centered on climate change activism in D.C.

For Nazar, animals first inspired her involvement in the fight for climate justice.

"When I learned about how species are being pushed to extinction due to climate change, I knew I had to take action," Nazar told *Teen Vogue*. A dedicated vegetarian, Nazar believes industrial animal agriculture is an overlooked aspect of the climate movement, explaining that the industry accounts for 14.5 percent of the greenhouse gas emissions in the atmosphere. "Corporations have corrupted our lifestyles and normalized so many toxic things that not only hurt our planet, but the people and wildlife living on this planet," she said.

Since helping to organize the D.C. Youth Climate Strike this year, Nazar has become the Maryland state lead for U.S. Youth Climate Strikes. First up was the Youth Climate Summit that took place from July 12 to 14 in Miami.

## 5. Isra Hirsi, 16

With a deep passion for changing the world, Isra Hirsi first got involved with climate justice in high school, when she had the opportunity to learn more about the earth and its environment. Since then, she has joined fellow teen environmentalist Haven Coleman in co-founding and co-executive directing the U.S. Youth Climate Strike this year. The organization's first demonstration was held in March and saw thousands of people around the world join in the movement.

"If we don't stop the climate crisis soon, those already impacted will be hit even more and generations like mine won't have a livable future," Hirsi told *Teen Vogue*.

Currently, Hirsi, whose mother is U.S. Rep. Ilhan Omar (D-Minn.), is helping U.S. Youth Climate Strike try to organize a climate debate to help educate of-age voters on presidential candidates' stances on climate change.

## 6. Alexandria Villaseñor, 14

At fourteen years old, Alexandria Villaseñor has already made a name for herself as a teenage environmentalist. Dedicated to climate activism, the teen is best known for her individual protests outside the U.N. on Fridays, when she skips school to sit on a park bench outside the building. Her strike began as a sign of solidarity with Greta Thunberg's call for school strikes in Europe.

Villaseñor is also involved with the U.S. Youth Climate Strike and is the founder of Earth Uprising, an organization dedicated to fighting climate change.

Villaseñor told *Teen Vogue* her activism began shortly after the devastating Camp Fire in California last year. At the time, she was visiting family in Northern California.

"I have asthma and, for my safety, my family had to send me back to New York City. When I returned, I began to research the connection between climate change and wildfires and learned that wildfires are burning longer and hotter because of climate change," Villaseñor explained. "I also saw Greta's COP24 speech, and I thought that world leaders would take action to reduce greenhouse gas emissions. When they didn't, I was upset, and it was on the last day of COP24, December 14, that I began my strike at the U.N."

This summer, Villaseñor is helping Earth Uprising launch the campaign #ClimateStrikeSummer, where participants spend eight Fridays striking outside various companies that need to address climate change.

### 7. Haven Coleman, 13

Sloths were the main motivator for Haven Coleman's dive into environmental activism. She told *Teen Vogue* that a fifth-grade lesson about deforestation was the catalyst for her commitment to ending climate change.

"I was so mad that we were making the world so sick and hurting so many people, killing so many people. I knew that it was not about saving the sloths, it was about saving all the organisms that live on it," Coleman told *Teen Vogue*. "I started educating myself, then my parents, then the kids at different schools by doing presentations; then I started talking to politicians, speaking at rallies and events. I was doing everything I could do."

Having participated in climate work since she was ten, Coleman's transition to founding her own climate-focused organization was only natural.

"Millions of people will be displaced, millions will starve, billions of plants, animals, and organisms will go extinct," she said. "So much pain and suffering for all the things living on this earth—all made by us. This is a fight that will determine life and death for so many; this is a fight that is worth fighting for"

## 8. Xiuhtezcatl Martinez, 19

The youth director of Earth Guardians and author of the book *We Rise*, Xiuhtezcatl Martinez is an Indigenous climate activist who is using his voice as both a hip-hop artist and environmentalist to guide a global, youth-led movement dedicated to protecting the earth. Martinez began his activism at six years old, speaking at climate summits around the world. He's even addressed the General Assembly at the U.N. and received the 2013 U.S. Community Service Award from President Barack Obama, all before age twenty.

Martinez is perhaps best known for his lawsuit against the U.S. government. In *Juliana v. U.S.*, first filed in 2015, he argued that the federal government failed to limit the effects of climate change. Martinez has also worked locally to remove pesticides from public parks, contain coal ash, and place moratoriums on fracking in his home state, Colorado. In 2020, Martinez will be able to vote in his first presidential election. In an op-ed for *Teen Vogue*, he explained why he'll be supporting Senator Bernie Sanders in the 2020 Democratic presidential primary.

## 9. Jayden Foytlin, 15

A member of *Teen Vogue*'s 21 Under 21 class of 2018, Jayden Foytlin has become a major player in the fight against global warming while still in high school. Foytlin, an Indigenous and Cajun teen, is one of twenty-one people (including Martinez) suing the federal government in *Juliana v. U.S.* Along with her mother, Foytlin is also protesting the Bayou Bridge Pipeline, which would carry oil from the Dakota Access Pipeline into Louisiana, Foytlin's home state.

# Greta Thunberg Wants You—Yes, You— to Join the Climate Strike

**LUCY DIAVOLO**

*September 16, 2019*

I'm on the subway headed to Manhattan to meet Greta Thunberg, the sixteen-year-old Swedish climate activist who pioneered the climate strike movement, and I'm absolutely kicking myself for forgetting my travel mug. The iced coffee I'm sipping is in a single-use plastic cup—straw and all—and here I am on my way to meet arguably the most visible climate activist in the world.

Having completed a transatlantic journey by sailboat, Greta is scheduled to speak at the United Nations General Assembly's Climate Action Summit, another chance she'll have to make her no-nonsense appeal to world leaders about the urgent necessity of international action on the climate crisis. She's famous for being ruthlessly frank with the global elite, so when I meet her in a midtown conference room on a Friday morning, I'm surprised to find a reserved young woman who speaks softly after carefully considering each question I ask.

What's less surprising is the steadfast confidence and grave seriousness that emanates from this teenager who has given voice to an entire generation's existential fear and energized a worldwide movement demanding everything necessary and possible to save our planet.

Asked about how she's liked her visit to the States since her August 28 arrival, she praises the "really nice" people. Other highlights: keeping up her routine of unwinding with long walks by strolling through Central Park and visiting New York's museums, including her (fitting) favorite, the American Museum of Natural History.

But she does have one note about the city that could apply to much of the United States: "You're obsessed with air conditioning."

## A Global Sensation Tackling a Global Threat

Greta is shy and serious in person. She considers questions and gives thoughtful answers. I see this during our Friday interview and again on Monday, September 9, when she's on stage with journalist and activist Naomi Klein at an event sponsored by *The Intercept*. The event spotlighted Greta in conversation with Klein, but also featured Xiuhtezcatl, Vic Barrett, and Xiye Bastida—all accomplished climate activists age twenty or under who each offered a vision of their life in 2029 were the Green New Deal to be enacted.

The line stretches down a city block to get into the auditorium at the New York Society for Ethical Culture. The pew-like seating and giant halo-shaped chandelier overhead lend a vague religious overtone to the evening. While some might think these youthful climate activists are preaching to the choir, at the moment, it feels more like they are speaking gospel.

In her introduction ahead of their talk, Klein calls Greta "one of the great truth-tellers of this or any time" and praises her bravery for calling out the world's rich and powerful to their faces at the World Economic Forum in Davos earlier in the year. Greta takes the stage to a standing ovation and talks with Klein about the climate strikes, her Asperger's diagnosis, and her boat trip to the States, which she says was lengthy but worth it for the sights of wildlife and the view of the stars she got out on the open ocean.

One on one, Greta comes off as skeptical of the attention she receives. But in front of a full house that prizes activist sensibilities, that skepticism makes her a dynamo. The crowd loves her for laughing off online trolls and for researching absurd conspiracy theories about herself.

## Fighting for the Future, Living in the Now

Greta is in a line of work that can be notoriously difficult: activists often struggle to support themselves in the long term, and the emotional toll of the work can be a serious burden to bear.

She's been open in the past about how she entered activism as she was coming out of a very serious period of depression where she wasn't eating or speaking. Just days before we sat down to chat, she had shared on Instagram how she's been bullied for having Asperger's syndrome, calling out "haters" and writing in a caption, "I have Asperger's syndrome and that means I'm sometimes a bit different from the norm. And—given the right circumstances—being different is a superpower."

Klein asks Greta about this, too—and Greta explains that the way her brain operates can empower her as a climate activist.

"Without [my Asperger's syndrome], I wouldn't have noticed this crisis," Greta says. She tells Klein that after recognizing the climate crisis for what it was, she felt she had no other choice but to take action. She sees that drive, and that of other climate activists she knows on the autism spectrum, as evidence that there's something about the condition that makes for good activists.

"I think it has something to do with how we walk the walk and we don't have the distance from what we know and what we say to what we do and how we act," Greta says. "And without my diagnosis, I also wouldn't have been such a nerd, and then I wouldn't have had the time and energy to look through the boring facts and still be interested."

Back in the midtown conference room on Friday, I ask Greta if she views her experience with depression as a potential source of a superpower, too, and she tells me how, even if being different can cause depression, she views it as a strength.

"Depression is something that often people who are different suffer from, either because they work too much or because they are being bullied because they are different or just because they don't feel right in this society—that they feel everything is meaningless," she says. "That is often the people who think a bit outside the box and who can see things from a different, new perspective."

"We need these people, especially now, when we need to change things and we can't see it just from where we are. We need to see it from a bigger perspective and from outside our current systems," she explains. "That's why people who are different are so necessary: because they contribute so much. Therefore, we need to really look after the people who are different and who may not be heard. We need to listen to those and to look after each other."

## Climate Change or System Change?

Greta's weekly strikes started outside her home country's parliament in August 2018. In those early days, it was just her. But she has become a focal point for a youth movement that is taking over the world. Just after our initial interview, Greta joined local climate strikers in New York for the second week in a row.

I ask Greta if she thinks people with economic security in richer countries have any special responsibility to the climate movement, and she tells me she does.

"We have to lead because we have already built infrastructure that other countries need to build, and it takes carbon dioxide to build that infrastructure and to make sure that people in poorer countries can be able to heighten their standard of living," she says. "We have to also give [poorer countries] the opportunity to adapt [to climate crisis]. Because, otherwise, it doesn't make any sense."

The issue of systemic change is on a lot of climate activists' minds. In 2017, the Carbon Majors Database compiled a list of the largest institutional sources of global greenhouse-gas emissions, finding that just twenty-five companies have accounted for more than half of global industrial emissions since 1988. Environmental groups have seized on these numbers as evidence of the systemic nature of carbon emissions.

While she clearly thinks our current political and economic system has exacerbated the crisis by its refusal to act, she's less willing to say whether she believes capitalist countries can make the changes required of them, telling me, "I don't think it will be up to us teenagers and children to actually solve the problem."

"We young people are building this up," Greta says, making it clear that the strikes are a message to world leaders. "They always say they have listened to us, so this is a chance for them to prove it."

Greta generally operates from a place of granting authority to science. Klein asks about her insistence that she's not prescribing action for political leaders, but just asking them to listen to the science. Greta replies that she vets her speeches with scientists—asking them to check not just for factual accuracy but also for misunderstandings or clarifications. Even if she's taking a year off school, she's clearly still doing her homework.

Klein is an expert on the climate crisis in her own right. The author of books like *This Changes Everything: Capitalism vs. The Climate* and *On Fire: The (Burning) Case for the Green New Deal,* she has been a powerful voice connecting our political and economic systems to the climate crisis. She tells me backstage before Monday's event that she believes the climate crisis is one outcome of the capitalist system, prompting me to consider the two side by side.

But the most political Greta gets during her conversation with Klein is when she's asked her about concerns within the Democratic Party that a climate response like the Green New Deal is too expensive.

"The money is there," Greta tells her. "If we can save the banks, we can save the world."

## The Next Global Climate Strike

Ahead of her speech at the UN Climate Action Summit on September 23, Greta is also trying to get the word out about the next global climate strike, set to begin on September 20, the latest major action in the climate movement's world-saving gambit. She hopes that more people than ever will take the day to strike from work or school.

She tells *Teen Vogue* that whether it's the Swedish government, major corporations, or the United Nations General Assembly, climate strikers are trying to generate political momentum to address the climate crisis—to push governments, corporations, and fellow citizens to do "what is required and what is possible."

"Please think about it from a bigger perspective," she says when I ask if she has advice for those unsure about joining the September 20 strike. "Not just from today, but imagine yourself in about twenty or thirty years. How do you want to look back at your life? Do you want to be able to say that you did fight against it and tried to push for a change early on? Or do you want to say that, 'No, I just went on going like everyone else because it was too uncomfortable.'"

"If you can't be in the strike, then, of course, you don't have to," she continues. "But I think if there is one day you should join, this is the day."

The fact that "perspective" came up twice during our interview doesn't surprise me, nor does it surprise me that Greta talks about the view of the stars from her sailboat or the way she views Asperger's as her superpower. Youth climate activists have a way of giving those of us who might be older and more jaded the perspective to see the potential for a future without crisis, but meeting Greta affirms that this is about more than just hijacking youthful optimism. It is about welcoming the perspective of a generation that is fighting for its own future—for the right to live.

# Behind the Scenes
# with the Climate Strike's
# Teen Organizers

**SARAH EMILY BAUM**

*September 20, 2019*

Teen Vogue *reporter Sarah Emily Baum spent weeks with the student organizers of the New York City Youth Climate Strike, attending planning meetings and getting up to speed on how they put together one of the largest mass mobilizations for climate justice in history.*

## August 21, 2019, About one month
## before the Youth Climate Strike

In a New York City basement, just down the street from Trump International Hotel, fifty kids sit in folding chairs trying to come up with a plan to save the world.

"We really want this organizing body not to focus on September 20 as a goal, rather as a catalyst," says Xiye Bastida, a seventeen-year-old Indigenous climate activist and member of the NYC Youth Climate Strike's core committee. Now a resident of Morningside Heights, in New York City, Xiye spent most of her life in San Pedro Tultepec, Mexico. Back then, she saw the flooding of her hometown as an anomaly, rather than a symptom of a much more sinister problem—the climate crisis.

On September 20, Xiye and activists around the world will mobilize for the Youth Climate Strike. At the helm of this action is Greta Thunberg, a sixteen-year-old Swedish environmentalist who went viral for going on a weekly school strike to protest lawmakers' inaction in response to the climate emergency. Thunberg will be joining the march in New York City ahead of the United Nations Climate Summit, where major world powers will decide whether meaningful steps will be taken to fight the climate crisis, or if they will continue stalling until it's too late.

Until then, students like Xiye are building the movement from the ground up. "[Smaller actions like this are] important because organizing is always the backbone of every big action," she says. "You need to be coordinated if you want to have a united front."

Although some global warming is organic, climate scientists say human activity (like the emission of fossil fuels) has caused temperatures to increase at an accelerated rate. When the temperature warms, glaciers melt, sea levels rise, and natural disasters increase in frequency and decimate the globe. According to the United Nations, a global warming of just 1.5 degrees Celsius (2.7 degrees Fahrenheit) would be catastrophic.

Young people—especially those who are Indigenous, lower-income, or people of color—are expected to be hit the hardest. That's, in part, why they're leading this movement. In addition to Greta and Xiye, youth-led organizations from around the country and around the world are driving the cause forward, each adding their own unique voice to the chorus of young people demanding change. Zero Hour, led by seventeen-year-old Seattle native Jamie Margolin, uplifts marginalized voices on the front lines of the climate fight; the Sunrise Movement, cofounded by twenty-six-year-old Varshini Prakash of Boston, prioritizes lobbying for policies like the Green New Deal; Extinction Rebellion Youth uses nonviolent direct action to bring a sense of urgency to the climate crisis. The list goes on.

"Everyone has the power to make our voices heard," Xiye tells *Teen Vogue.* "Now, we have the responsibility to make our voices heard."

At the end of the meeting, the students head outside. Some are in large, blow-up dinosaur costumes—"because we'll be fossils too, if we don't act"—and in the streets, against the backdrop of Central Park and traffic lights, they dance and sing their protest songs.

## August 28, 2019, Three weeks before the Youth Climate Strike

After spending fifteen days on a zero-emissions yacht sailing across the Atlantic Ocean, Greta Thunberg sets foot in New York City for the first time. She is greeted by a swarm of press and a crowd that includes Xiye and a hoard of other young activists who are chanting her name and anticipating the upcoming Friday strike—Thunberg's first in the United States. They plan to convene outside the United Nations headquarters in Manhattan later that day.

By this point, eighth grader Alexandria Villaseñor, a fourteen-year-old New York City student, had been striking there every Friday for thirty-eight consecutive weeks. Alexandria, who has asthma, began her climate activism after a family visit to California left her ill and unable to breathe. The smoke from rampant forest fires had seeped into their house.

"I saw the connection between that and climate change," says Alexandria. She founded Earth Uprising, an organization promoting climate literacy among young people, and is serving as a core organizer of the NYC Youth Climate Strike. She has continued her strikes through rain, cold, and a polar vortex, as well as a death threats from right-wing trolls and climate deniers.

Nonetheless, Alexandria says, "I wanted to defend my hometown." That's why she made a simple white sign and started her strike on a little metal bench outside the United Nations, where she often sat alone.

This was not one of those weeks.

Hundreds of students, activists, and journalists descended upon her quiet little bench. A circle of teenagers stood outside, and someone "shoved" Alexandria to the middle, right next to Greta. In the chaos, it took almost thirty minutes just to cross the street, Alexandria tells *Teen Vogue*.

"Halfway through at the park we got the message the president of the U.N. General Assembly wanted to meet with us, and they shuffled us back to the entrance of the building," Alexandria says. The teen girls—Alexandria, Greta, and Xiye—told her that they were fighting for changes, like a more ambitious international greenhouse gas emissions agreement.

"We are out of time," Alexandria tells *Teen Vogue*. "We're in a crisis. But now we're raising awareness and getting attention from people in power—and we're just getting started."

## September 6, 2019, Two weeks until the Youth Climate Strike

"I sure wasn't doing that when I was their age," a photographer says. On the sidewalk, kids are making their protest signs. Amanda Cabrera, an eight-year-old climate striker who lives in New York City, was writing climate justice slogans on her arms, but needed help from a friend because she wasn't quite sure how to spell "planet."

"[The climate crisis] makes me feel angry because Earth is giving us life but we aren't giving it life," Amanda tells *Teen Vogue*. "We need the planet but the planet doesn't need us."

She attends the strike that week with her friend, nine-year-old Aaron Thomases. "[I've been an activist] since the COP24, when my mom showed me the video of Greta Thunberg speaking," Aaron says. "I want other kids to know [the climate emergency] is real, it's happening, and if you want to be a superhero, now is your chance."

After an impromptu press conference next to Alexandria's bench, where Greta fielded questions from reporters in multiple languages, the group moves to a nearby park. Even in New York, we can feel ripples from Hurricane Dorian, a Category 5 storm that decimated the Bahamas. Wind blows signs across the street and inverts umbrellas in the rain. Still, the strikers hold strong.

One student protestor talks about going to school for weeks without power after Hurricane Sandy in 2012. Another, eighteen-year-old Shiv Soin, tells the crowd he had gotten ill from drought and sandstorms while visiting India, and he later realized this was yet another symptom of the climate emergency. And Clarabell Moses, a teen activist visiting New York from California, talked about her hometown of Paradise— the very same town where wildfires had given Alexandria Villaseñor an asthma attack and also launched Clarabell's fight for climate justice.

"My entire childhood home burned to the ground," she says. "I watched the people around me lose their homes, loved ones, and even their lives. Civil disobedience like this is important because when we disrupt the system, we break it down and build it back up better than before."

"At the forefront of every revolution has been young people," adds Ayisha Sidiqqa, a New York City student. "From the Greensboro sit-ins to the anti-Vietnam War protests, to the movement happening right now in Hong Kong—when students speak together, the whole world listens."

**September 19, 2019, One day before the Youth Climate Strike**
The Amazon rainforest is burning. Time is running out.

According to a recent count, more than ninety-three thousand fires were decimating the Amazon—burning more than one soccer field per minute, according to the INPE, Brazil's National Institute for Space Research.

Forest fires are in fact worsened by the climate crisis, but these fires in particular are man-made. And the government is doing little to stop them.

As kids around the world gear up for the Youth Climate Strike Friday, Brazil's environmental minister awaits a meeting with climate change deniers, and the president of the country actively fights international efforts to extinguish the blaze.

"It didn't catch fire, it was set on fire by [those] who want to burn it down for agriculture and to make money," seventeen-year-old climate activist Jamie Margolin, founder of Zero Hour, tells *Teen Vogue* ahead of the September 19 meeting. The Washington State native has also participated in the school strikes—and hers specifically focus on the Amazon rainforest, often called "the lungs of the planet."

"If we lose that, there's no hope," Jamie says.

The Amazon rainforest is a carbon sink. This means it drains carbon dioxide, the primary greenhouse gas associated with global warming, from the atmosphere. The gas accumulates in trees over time, meaning that when the Amazon burns those gases are suddenly released back into the atmosphere.

"It's a war on nature," Jamie says. "It's very targeted, purposeful violence against . . . our Earth."

She also says fighting the climate emergency means fighting the root causes: excessive capitalism, colonialism, and racism. It's a sentiment not lost on her friend and fellow activist David, 19, of Medellín, Colombia, who is on the front lines of this fight. He asked that his full name not be used because he says environmental activists like him are frequently targeted by paramilitary groups and guerrilla forces in Latin America.

"From the windows of my house, I can see the mountains burn," David tells *Teen Vogue*. This past year, Colombia was ravaged by fires, flooding, and crop failure, yet another symptom of the climate emergency.

"[The fires in the Amazon] make me feel like this strike is more necessary than ever," Jamie says. "While mass public opinion calls for climate action, those at the top profit from mass destruction."

Still, Jamie is hopeful. "There's only so long those in power can get away with going against the common good and the will of the people," she says. "The tables will turn starting on September 20."

# I Protested for the Green New Deal at Mitch McConnell's Office. Here's Why.

DESTINE GRIGSBY

*February 27, 2019*

I never imagined I would sleep on the sidewalk outside my senator's office. But on February 21 that is exactly what I and ten other people did. We were there to demand that my senator, Mitch McConnell (R-KY), come, look us in the eyes, and listen to the young Kentuckians who are demanding a Green New Deal. Sadly, it was easier to imagine McConnell ignoring us than listening to us.

In reality, far too many people in my state do not have clean water, suffer from black lung disease, and feel they have no economic future. Many of these public health hazards are linked to our state's relationship with the fossil fuel industry, which has polluted our environment and people's bodies, and left many with no options after a lot of jobs the industry created have dried up.

Despite the turmoil the fossil fuel industry has caused in Kentucky, McConnell has continued to accept a combined millions of dollars from the industry during his time in the Senate. I wonder if it weighs on him that some of his constituents go without clean water, clean air, and sustainable jobs, while he continues to collect money for his office. If McConnell cares about Kentucky coal miners so much, why has he failed year after year to pass federal bills that would repair coal mines, support disability for those with black lung disease, and ensure that miners have pensions? Why did he try to sell us on reducing EPA limitations on coal when analysts said these rollbacks would have little positive economic impact, which was McConnell's reason for supporting them?

Last week, I slept outside McConnell's office because Kentuckians shouldn't suffer. I have seen how climate change and fossil fuel industries hurt people in my home state. Across Kentucky, from coal ash spills in eastern Kentucky to the urban heat island effect in Louisville, people have been used as pawns by politicians and the fossil fuel companies that line their pockets. Kentucky needs a Green New Deal to shift our state's economy away from coal, ensure everyone the right to clean air and water, and retain the next generation of workers who are now suffering due to a lack of jobs.

I knew that President Donald Trump's decision to pull out of the Paris Climate Agreement in June 2017 would bring nothing but darkness to my state. It was at that moment, at sixteen years old, that I joined the Sunrise Movement, a group of young people organizing to make climate change an urgent political priority by planning a rally in Louisville to protest Trump's decision. I had never done anything like that before. I rushed to mobilize young people in my city, and together we held one of the first rallies led solely by high schoolers that I had ever seen.

On the day of the rally, it was pouring rain, but more than thirty young people were standing strong outside Metro Hall. We were united, singing, all holding up painted fists. We each wore a red bandana reading "For Our Future." It may not have been the most organized rally, but it certainly had energy: we marched, chanted, and sang for hours in the rain. Three local news stations covered our actions, and the city got to see that young people are ready to fight. For the first time, I understood that despite being high schoolers, despite not being old enough to vote, young people had the power to change the world.

I didn't know then that our rally would inspire young Kentuckians to keep mobilizing. But last December, more than seventy-five of us joined the Sunrise Movement in Washington, D.C., to support the Green New Deal, attracting the attention of national journalists. Within a week of making calls and posts and sending texts, we were able to fill up a bus of students to share their stories in D.C.; in fact, Kentucky brought more people to the action than any other state in attendance.

For us, the fight against climate change is common sense. As Kentuckians, we know that a Green New Deal can create millions of good jobs guarantee clean water, clean air, and a hopeful future for our impoverished state. Despite our conservative government and the struggles we face, Kentuckians have shown up en masse, ready to fight.

A few weeks ago, Senator McConnell announced that he would put the Green New Deal resolution to a vote in the Senate. This is not because he supports the resolution—he's rushing it to a vote so that he can crush our momentum.

So, on Monday, February 18, we went to McConnell's Louisville office and demanded a meeting with him, that he meet us face-to-face and admit that he thinks the millions from fossil fuel industries are more important than my generation's future. We were denied a meeting with McConnell and physically blocked from entering his actual office. I was infuriated. This was during Senate recess, but he didn't even have the decency to speak with young Kentuckians who showed up at his door, the same constituents who will suffer the most from climate change.

Little did he know, that was just the beginning.

Every day of that week we stood outside McConnell's office demanding a meeting, showing him that we will no longer be ignored. On Wednesday, young people with the Kentucky Student Environmental Coalition went to his office in Frankfort; Thursday was the day I camped outside overnight with ten others, where it was terribly cold and uncomfortable sleeping on concrete; on Friday, I walked out of school and rallied with students who had walked out of high schools and college classes at the University of Louisville.

On Monday, February 25, a group of fifteen young Kentuckians traveled to Washington, D.C., and again demanded that McConnell listen to us.

In the capital, hundreds of young people joined us in the halls of McConnell's office building on Capitol Hill. Cameras from cable news stations across the country also crowded the hall outside McConnell's office. I led the group into the office carrying a box containing more than one hundred thousand signatures from people around the country who want a Green New Deal. According to a recent poll, 80 percent of the public supports most of the measures in the resolution.

Though we met with his office's Kentucky director, we still wanted to face Senator McConnell in person. When I heard he wasn't in his office, I wasn't surprised. I think McConnell doesn't have the courage to admit that he prioritizes the rich over our lives. Stories of pain and suffering in Kentucky, especially from kids, often go unheard. But that day, in his office, many Sunrise members were moved to tears as we shared our stories to show everyone why the Green New Deal is the only way to fight climate change, poverty, and pollution in Kentucky.

My adrenaline was pumping. Bright camera lights beamed down on us. Police flooded the hallway, warning that they would arrest us. After an hour of sharing stories and singing songs in McConnell's office, forty-two young people were taken into custody and, for the first time in a while, I felt hopeful. People I didn't know were willing to get arrested to stand up for me and my state. This was the first time I understood that Kentucky was not alone in this fight.

Afterward, those of us who hadn't been arrested rallied outside. I listened to a fellow classmate, Oli, tell a story about why she is fighting for the Green New Deal. She talked about her family in the Philippines and those who can't fight for themselves. She talked about how scared she is at the thought of her mother and father not being physically able to escape the next typhoon, which may pummel the Philippines harder now because of warmer ocean waters. When I first met Oli, she was shy and afraid of public speaking. I almost didn't recognize the person in front of me raising her voice outside McConnell's office, through tears, demanding he protect her family.

These rallies aren't just about chanting and being on the news. They are about us defending our right to be heard and our right to a home, to clean air and water, and to a livable future. Listening to Oli tell such an emotional and dire story cut through the noise and brought me back to earth. She reminded me that if we let McConnell play games with our future and do nothing, we lose. We all lose. But, if we fight back, if we share our stories, we can build a bigger movement and win.

# Five Youth-Led
# Climate Justice Groups
# Helping to Save the Environment

MAIA WIKLER

*March 28, 2019*

We've seen youth rising to the call and become climate activists over the last year, and that's largely because the stakes have never been higher.

Young people around the world are demanding urgent action to address the ongoing climate change crisis, and are ramping up efforts. Greta Thunberg's call to action at the United Nations climate change summit, which led to a Nobel Peace Prize nomination, Sunrise pushing forward the Green New Deal, and the ongoing school strikes around the world—including one held on March 15—have continued to show this generation's commitment to the cause.

These visionary leaders are creating radical change, fast. Hope not only exists in their bold and unapologetic approach but also in the sheer size of this generation. The Pew Research Center projected that millennials are set to surpass baby boomers in population size in 2019 and represent nearly 40 percent of the electorate by 2020, according to Center for American Progress.

To honor their efforts, *Teen Vogue* spoke with five youth-led climate justice groups for a glimpse into the many ways young people can get involved.

## 1. SustainUS

While politicians congratulate themselves for addressing the climate crisis with the Paris Climate Agreement, the fossil industry is still able to wield access and power in the negotiating spaces at the U.N. Through symbolic and direct actions, SustainUS brings youth to international negotiations to dismantle the political elite's narrative and demand stronger, urgent action.

Most recently, SustainUS sent youth delegates to the World Bank meetings in Bali, Indonesia, to speak out on fossil fuel corruption. In 2017, the group organized an action that went viral when youth delegates disrupted the White House panel promoting fossil fuels at the U.N. climate change conference in Bonn, Germany.

Daniel Jubelirer, COP24 delegation leader for the annual United Nations climate conference, spoke to *Teen Vogue* about the key strategy and role of SustainUS. "We wield storytelling as a weapon against complacency. We are bringing young people who have a lot of lived experience with injustice from climate impacts and racial injustice," he said.

Phillip Brown, a twenty-year-old queer immigrant from Jamaica and SustainUS COP24 youth delegate tells *Teen Vogue*, "My presence there was a tangible form of reparation in the sense that Black and brown people don't have the resources to make it into these spaces, even though we are some of the most impacted by climate crises. By showing up in these international spaces, I am reclaiming what's been taken from us for centuries, our right to take up space and voice our demands for solutions that center the needs of our most vulnerable communities."

## 2. Those leading lawsuits in defense of the climate

Youth around the world are using litigation to hold governments accountable to stop climate change and address environmental injustice.

In the U.S., twenty-one young plaintiffs are suing the U.S. government for violating constitutional rights to life, liberty, and property by allowing and promoting the use of fossil fuels despite knowing they are directly causing the climate crisis.

In Canada, youth recently launched a class-action lawsuit in Quebec arguing that the government is violating the rights of young people by failing to take urgent climate action.

"People under thirty-five will be most affected, we will be here to experience the worst impacts of climate change. It might take ten years to get a final decision from the court but this lawsuit sends a clear signal to all governments that they need to take climate change seriously," Catherine Gauthier, executive director of ENvironnement JEUnesse and the representative plaintiff in the case, tells *Teen Vogue*.

## 3. Uplift

Vast areas of the Southwest have been dubbed "energy sacrifice zones," which means millions of acres of federal land are being used and polluted for energy extraction. The Southwest is already experiencing some of the most dire impacts of climate change with massive heat waves, megadroughts, and rapidly diminishing water sources. Uplift, an award-winning collective of youth grassroots leaders, is tackling this ongoing crisis.

Brooke Larsen, executive coordinator of Uplift, tells *Teen Vogue* that the organization is unique because they center voices from the front lines. "We take on a radical stance," she says, focusing on colonialism and capitalism as the group builds alliances with different groups in the struggle against climate change.

Uplift strives to be a connective force for the region by organizing an annual three-day outdoor climate convergence on the Colorado Plateau, training youth leaders in grassroots organizing skills, and using storytelling to amplify marginalized voices across the Southwest. Georgie, a young Hopi woman and Uplift organizer, tells *Teen Vogue* that her community's core values are "from the Earth in reciprocity, respect."

"I grew up with that way of living. Here in Hopi, we have Peabody Coal. There is mining, oil, and gas all in our backyards, on our sacred lands. Uplift creates a political space that brings people together from all over the Southwest, connected by the fact that we are all affected one way or another."

## 4. University divestment campaign organizers
The student-led divestment movement is putting the pressure on academic institutions around the world to uphold their commitment for the interest of the public and greater good by cutting their financial ties with the fossil fuel industry, which they argue is reckless in the face of climate change.

According to *Vice*, as of May 2018, 133 schools, including Stanford, Oxford, Cambridge, and Yale, had divested from fossil fuels since the movement began in 2011. Today, educational institutions with previous investments in fossil fuels valuing over $1 trillion have committed to divestment from these industries because of these student-led divestment campaigns.

The success of this movement, however, is not only in the number of divested institutions or the amount of money moved. Just ask Emilia Belliveau, who has a master's in political ecology and is a former divestment organizer at Dalhousie University and spent three years researching and interviewing the organizers of the movement for fossil fuel divestment on campuses. She tells *Teen Vogue*, "That perspective doesn't acknowledge the social impacts of fossil fuel divestment as a movement. This movement has empowered thousands of young people around the world to be skilled community organizers with an understanding of climate change that challenges systemic power and inequity."

"[Universities are] still engaging in colonialism in this era of reconciliation," Sadie-Phoenix, two-spirit grassroots organizer and community advocate who led the divestment campaign at the University of Winnipeg, tells *Teen Vogue*. "Educational institutions have a responsibility to move forward with reconciliation after the history of residential schools. It can't do that when it's actively colonizing by failing to address climate change and threatening land and water. Infrastructure that's gold LED standard is greenwashing when it's funded by oil companies. Divestment is a way to uphold reconciliation."

To learn more about starting a divestment campaign at your school, go to DivestEd.

## 5. Sunrise Movement

Sunrise Movement is redefining youth activism in the U.S. with the meteoric rise of their movement to make supporting the Green New Deal a mainstream position. Teaming up with the youngest woman ever elected to Congress, Alexandria Ocasio-Cortez, Sunrise Movement and the Green New Deal want to transition the U.S. to 100 percent renewable energy by 2030.

Varshini Prakash, cofounder of Sunrise Movement, tells *Teen Vogue* that the group is led by young people. "Everyone who launched Sunrise Movement was under twenty-six at that time. For the first year before we launched, it was just a few of us—about twelve people—and we had no idea that it would become this large of a movement."

Now, nearly two years later, Sunrise Movement boasts thousands of members and trained youth activists across the U.S. "We are combining protest organizing and electoral organizing together into one strategy, which is pretty unique, as opposed to many other groups who talk about it from the perspective of what we can get from our existing political reality," Varshini says.

"The millennial generation is not starting from a place of what is politically feasible in this moment; youth are pushing to stretch the imagination of what is possible."

# Teens Are Suing the U.S. Government Over Climate Change:

## The Trump Administration Is Trying to Stop Them.

**ROSALIE CHAN**

*August 7, 2017*

Jaime Lynn Butler, at the time a rising junior at Colorado Rocky Mountain School in Carbondale, Colorado, tells *Teen Vogue* that she got her first taste for activism from her family. Growing up on the Navajo Nation reservation in Arizona, they'd joined their tribe in defending its water rights.

Now, Jaime is one of twenty-one young people who are suing the federal government for promoting measures that harm the environment, like allowing fossil fuel extraction, which has caused carbon emissions to "dangerously increase," and subsidizing the fossil fuel industry. The group is made up of people ages nine to twenty-one, and they hope to claim what they say is a constitutional right to a healthy climate. They first filed the lawsuit in 2015, during the Obama administration.

On June 28, U.S. magistrate judge Thomas Coffin set the official trial date for February 5, 2018. But on July 25, the U.S. Court of Appeals for the Ninth Circuit put a temporary pause on the case, as the court is considering a petition from Trump administration.

The petition that the Trump administration filed requests a rarely used procedure called writ of mandamus, which, according to the *Washington Post*, basically lets higher courts review—and even possibly overturn—the decisions that lower courts made before the trial even happens. The petition is asking to overturn an earlier decision that denied the administration's request to dismiss the lawsuit. "It's a very rare type of motion that can be made at the court of appeals or Supreme Court asking the appellate court to require the district court judge to do something different from what it has done with an ongoing case," Julia Olson, the executive director and chief legal counsel of Our Children's Trust, tells *Teen Vogue*. "The only time that usually happens is if the court does something egregiously wrong."

The teens who are part of this lawsuit are working hard to prepare, even as the federal government tries to stop them. Jaime is motivated by the issues she's seen on her reservation. "Because of droughts . . . it's been so hard for a lot of people to raise livestock to keep everything alive," Jaime says. "There's a lot of elderly people . . . that live really far in the desert without electricity or running water, and it's really hard for them. If this keeps going, there won't be a lot of water."

The young people allege that the government has known about the dangers of climate change but has not done nearly enough to reduce the emissions causing climate change. If the lawsuit moves to trial and the court rules in the youths' favor, the court must set a safe standard for carbon dioxide in the atmosphere, and that standard would be enforceable. It also asks the court to order the federal government to set a plan for limiting emissions.

"We are arguing that the government has known about climate change for fifty years. We see that and have evidence for it," Aji Piper, a sixteen-year-old rising senior at West Seattle High School at the time of writing, tells *Teen Vogue*. "Because the government has violated our rights for so long, they now need to be held accountable and take responsibility and reduce the effects of climate change and reduce our emissions."

Aji has also been involved in a state lawsuit in which young people, ages twelve to sixteen at the time, sued the State of Washington because its clean-air rule didn't do enough to reduce emissions. Last December, a Seattle judge ruled that the youths can move forward with the case against the state.

To prepare for the federal trial, Aji has been studying scientific papers to make sure he's totally knowledgeable, as "[Trump] has made it so people are paying more attention to issues they care about, especially in a time when things are being rolled back and people are really upset about it," he says.

Nathan Baring, a seventeen-year-old rising senior at West Valley High School, at the time, in Fairbanks, Alaska, agrees the case has gained more momentum since Trump became president, especially as Trump has enacted measures that work against effective climate change action. But now, Nathan says, the government is doing everything it can to stall the case.

"I've lived in Alaska my entire life," Nathan says. "Just in my local area in Fairbanks, one of the most vivid signs of climate change is the frequency of winter rain." Like Jaime, Nathan has been involved in local environmental activism. Fairbanks has some of the worst air pollution in the country, and he does not want to grow up in an area where he can get respiratory issues from playing sports outside. Other parts of Alaska are also facing the direct impact of climate change.

"The biggest issues that are current to Alaska are certainly permafrost melt, coastal erosion, melting sea ice, and wildfires," Nathan says.

The legal team is working to prepare expert reports for the trial, bringing in evidence on the impact of climate change. Some of the teens on the case are also helping, conducting factual research and creating charts and graphs for the trial. During the week between the science march and the climate change march, some of them did a speak-out in front of the U.S. Supreme Court building.

Climate change affects everyone around the world but will especially hurt young people who have to grow up with its effects. Jaime notes that it's also important to recognize that people must fight climate change through an intersectional lens, as the effects of climate change hurt people of color the most.

"A lot of other reservations are having a hard time with water," she says. "There are so many things that have happened on a lot of reservations where the government has taken away or ruined some land . . . on our reservation. If they tried to help with climate change and try to stop it, I think it would be really helpful for not just the Navajo reservation but other reservations."

As these teens wait for the court to consider the petition, they hope to gain support across the country for the trial. They will be planning solidarity rallies and days of action as the trial date approaches. For more information on the movement, visit youthvgov.org.

# How to Take Direct Action
# on the Climate Crisis
# at Your School This Year

*TEEN VOGUE* STAFF

*August 13, 2019*

The back-to-school checklist is usually pretty standard: ordering textbooks, buying new notebooks and pens, looking up the locations of your new classrooms. But this year young climate activists hope you add another item to your list: figuring out how to help address the impending threat of the climate crisis.

From pushing your administration to institute a recycling or compost program to participating in national walkouts, there are many ways to take direct action to make schools greener and stand up to defend the rapidly warming planet and those threatened by the changes.

We reached out to three different climate justice organizations—Sunrise Movement, Zero Hour, and the U.S. Youth Climate Strike—

124

and surveyed five different climate activists under twenty to hear what they had to say about advocating for climate justice in their schools. Zero Hour's deputy communications director Natalie Sweet, 16, and Georgia co-executive director Zeena Gasim Abdulkarim, 18, told us how they launched initiatives in their schools. The U.S. Youth Climate Strike's cofounder Isra Hirsi, 16, and activist Sabirah Mahmud, 16, explained what it's like organizing with classmates. And Sunrise Movement's pre-college organizing lead Rose Strauss, 19, shared why putting pressure on institutions like schools is so important.

Below, read what they had to say to *Teen Vogue:*

## What kind of climate advocacy have you done at your school?

**Zeena:** I encouraged my classmates to practice sustainability in their lives and to take action against the progression of climate change by being conscious of their carbon footprints and how they could positively impact the climate movement. I did this through a politically and socially oriented club that I founded at my high school in my junior year. Through this collective effort, we successfully swayed our school's administration to make the shift from a nonrecycling establishment to a recycling establishment.

**Rose:** In 2017 and 2018, twenty people died in a mudslide, and wildfires burned down peoples' homes in my local community of Santa Barbara. One of my university's departments put out a report about the economic benefits of a fossil fuel project in our community—while taking money from fossil fuel corporations—so we confronted the professor who directs it, at his office, in May. It was part of our larger campaign of holding our school accountable for not holding to its values and mission. (UC Santa Barbara did not return *Teen Vogue's* request for comment.)

**Natalie:** I have started several climate initiatives at my school, such as the first high school chapter of Food and Water Watch's Take Back the Tap project, which aims to eliminate plastic bottled water at schools and universities. I also have written a climate action–focused op-ed for my school newspaper and urged my school community to participate in the climate strikes.

**Sabirah:** I have spoken with school administration about the need for more climate-friendly solutions in our school environment. Instead of giving plastic-heavy packaging with our breakfast, [I've urged them to] invest in more eco-friendly alternatives, such as juice boxes that don't need plastic straws.

I have also [urged] the administration to help our students start striking and skipping school on global strike dates, such as March 15 and May 3. To prepare everyone in my school environment for these strikes, I visited many advisories (homerooms) in the morning and gave mini-speeches to the teachers and students about why it's so important to come out and make sure that you['re] presen[t].

**Isra:** I used to be a part of my school's environmental club where we would organize [about] how to compost and recycle. Also helped the student walkouts in my school for the strikes by boosting on social media and passing out flyers.

## Why was it important to you to press for these issues?

**Rose:** We took action because our schools, our universities, are meant to prepare us for our future. But instead they are funding the same corporations that are actively destroying our chance at a livable future. People are dying from the climate crisis on our doorstep. Schools, especially ones on the front lines of disasters, must stand by young people in our fight to stop the climate crisis.

**Natalie:** Climate change is going to affect my generation and the generations to come the most; however, many people still do not know or recognize this threat. It is crucial for me and other climate activists to provide outlets for education and action in our communities.

**Sabirah:** Ten years from now I will only be twenty-six years old. Those in office or who have the privilege to make decisions about MY future are not making decisions that will give me the future that I, along with everyone in my generation, deserve.

My family in Bangladesh are losing their lives due to climate change and we're just sitting here in our privilege not doing anything because we don't see these people who are suffering.

## What was it like organizing fellow students?

**Rose:** There was so much energy. People were scared but also excited about having the power to make a change. When adults say no to us when we ask them to protect us, we know that we don't have to turn around and walk away. We can fight back through nonviolent direct action. We can demand that these adults—and the schools run by them—choose our long-term futures over short-term profits.

**Sabirah:** Organizing with fellow students is honestly so stress relieving because, if I'm going to be honest, adults are a little intimidating. I have made a lot of friends through organizing, and they're super kind, so that is probably the best thing about organizing with other students. They get all your problems with being an organizing activist and student because they experience the same thing.

**Isra:** It's a little harder getting peers to come to the walkouts/events because everyone is becoming very apolitical. Being one of the only outspoken students at my school, it has made it more difficult to get others to join me.

### Did you encounter any challenges in your work at school?

**Zeena:** My high school's administration was wary of the idea of transition[ing] to a recycling establishment because they believed they did not have the proper resources to collect the recyclable items from each classroom, so club members and I brainstormed until we established an effective system of schoolwide recycling collection. When we presented them with our proposal, they deemed it feasible, and we shortly began recycling as a school.

**Rose:** We weren't sure whether a confrontation or a sit-in would be best, so we brought food and everything in preparation for a night-long sit-in. People were also scared of the police coming. Scared about how we would look to the administration and staff at our school. Many people in my school and in the community told us not to do this. People are always worried to do direct confrontation. But if there is anything I have learned through organizing with Sunrise Movement, it is that direct confrontation changes what's politically possible.

**Natalie:** Most students would ignore my emails or refuse to see the purpose in taking steps to reduce climate change. This was slightly discouraging, but it just motivated me to work harder to spread the message of climate justice.

**Sabirah:** Many people don't really understand what we're doing, and some teachers have criticized me and said "strike on the weekend." Their words, however, don't really get through to me and I stay resilient. Students, at times, have criticized me about striking, usually by saying "striking doesn't even do anything" or making mindless comments to irritate me. I usually try my best to calmly explain that, "Hey, striking is actually important because it's expressing our First Amendment [right], and the system needs to know that we will risk our education to protect our future."

## What should parents, teachers, and administrators know about how to support youth climate activists in school?

**Zeena:** The best way older generations could possibly support younger generations in the climate movement is to educate themselves on the climate crisis, practice sustainability in their daily lives, attempt to influence their peers by having discussions related to the climate crisis, steer clear of supporting the companies that fund the fossil fuel industry, and most importantly, by voting.

**Rose:** When we go on strike for a Green New Deal, don't just applaud us, strike with us. When we make demands of our school and community, be vocal advocates for our work. School administrations and teachers: Stop shying away from taking stances because of partisanship. That is no excuse. This is not about Right or Left—this is about moving forward for the future of your students.

**Natalie:** Support the youth climate strikes! Listen to our voices and take part in this fight with us. Many youths can't vote yet, and votes from parents, teachers, and administrators for politicians who support climate action are extremely important and a major way to support youth climate activists. In the end, we need change on a broad, national level, and electing officials who will write laws for change is the way to address the climate crisis.

**Sabirah:** Parents can continue to support their children to strike and help them get involved with more climate initiatives to show our representatives that the youth are here. Teachers and school administrators should encourage their students to strike and make sure that they are educated about the severity of the climate crisis (because we need to learn about it!). Encouraging students to strike by not punishing them, or not giving them tests on the day of [a scheduled] strike, [is also] so important.

# Two Generations of Climate Activists Dish about Making Powerful People Uncomfortable

**ALLEGRA KIRKLAND**

*September 27, 2019*

Friday, September 20, 2019, saw what is estimated to be the biggest climate strike ever, with some four million people turning out in dozens of countries and all fifty U.S. states.

The Global Youth Climate Strike was historic, but the organizers behind it say they're just getting started. And indeed, we saw hundreds take to the streets in Brooklyn over the weekend, for a mass protest by frontline communities; on Monday, September 23, Extinction Rebellion blocked streets in Washington, D.C.; and a mass global climate strike is scheduled for Friday, September 27. This constant drumbeat of activity is intended to keep the climate crisis at the forefront of the news cycle and of peoples' minds, activists say.

At the head of this movement is a new cohort of teen organizers bringing a fresh sense of optimism and fire to the cause of climate justice. They're joining forces with an older guard of climate activists who have been doing this important work for decades.

This week, *Teen Vogue* hosted Fridays for Future activist Xiye Bastida, an Indigenous young person, Future Coalition executive director Katie Eder, and Greenpeace International executive director Jennifer Morgan at our office in lower Manhattan. The trio had a wide-ranging conversation about what generations of climate activists can learn from one other and why things feel different this time around.

*This conversation has been significantly condensed and edited for clarity.*

**Katie Eder:** What excites me about what happened with the 20th is that [the strikes on] March 15 [and in] May, that was kind of the youth wave—this new youth movement was over here, and this older movement that's existed for a long time is over here. But with September 20 we kind of combined the two, and I think that collective power, that collaboration, is what's allowing us to put down this new foundation that is the climate movement—that we can come together intergenerationally and intersectionally. And also, it's these big institutional groups with the grassroots; it's traditional environmental organizations with social justice groups.

**Jennifer Morgan:** It's way more transformational as well. You've managed to tap into this thing that connects all of us and make room for it, which is really generous and really smart. I think the [strike] before in 2014 was also exciting. I don't think the climate movement believed we could put people in the streets on climate change then. So people were really nervous about even trying. So it was so huge that there were the [four hundred thousand people in New York] back then. And now to see it really build, and to be able to see how it's just exploding in a way; you're building on it but continuing it. I can just tell you that I'm smiling because I've been in these conversations where people say, "Well, this is just a one-off right?" Or the other one that I love is, "You can control this, right?"

[*Everyone laughs.*]

**JM:** I'm like, "No and no."

**Xiye Bastida:** What we bring in the youth movement is that we're really tapping into every sector of society. We have youth in education reform, in policy work, suing the government, the consistency of striking every Friday. . . . As the climate crisis gets worse, we're going to get more activists until we stop it and we change it.

**KE:** From the beginning, we said, "Our goal is chaos. We want this to be as chaotic as it can be." Because if one person or one group or one organization is at the center and knows what's happening, it can't grow to the largest extent. But the fact that it's so grassroots, it's so decentralized, it's coming from this power of the local level—I think that's what sets it apart from other things. There's no hub; there's no circle that things are revolving around.

I think it's so much of a trust thing. It's trusting the businesses to self-organize, trusting the unions to self-organize, and I think putting trust in each other. There's a real culture shift that's starting to happen, especially around the youth movement, around the community that we're building, of really saying we're on the same team. No matter what organization or what sector of society you're coming from, we're kind of uniting around a common enemy.

**XB:** It's about collaboration and not competition. And it's so different from what we've been taught—we've been taught to be individualistic and to strive for personal success. But to see that we're all in this together in a collaborative way, and seeing that from the youth movement, [where] we make all our decisions through consensus, there's no hierarchy. We all love each other and respect each other. And in terms of the businesses, I always saw businesses as like a wall, and you couldn't get through, right? You couldn't talk to the people inside? But to see that so many businesses were supporting the strike, were shutting down, websites were shutting down, [thousands of] Amazon workers walked out; to see that businesses have that autonomy because we're creating that space for them and we are saying as youth we're gonna support you more if your business does that.

**JM:** I think one of the things that's made Greenpeace what it is, and [that] your movement [can do] as well, is bring a sense of unpredictability. I think what makes especially CEOs or heads of state a bit nervous is when they don't know what's going to come. It makes them uncomfortable; they need to be uncomfortable right now. And if they don't know where you are coming next, it keeps them more on notice than if they're like, "Okay, this is the next big mobilization, and the next one."

**KE:** That's why I think it's so amazing that—while we are working in coalition and we are working from one vision and one message—that we have different organizations that are working off different theories of change and tactics. So it does feel like from the local level, different cities and different organizations, there are constantly different types of actions happening at different times, and you never really do know. We might know 'cause we're living it every day, but you never really do know—the general public, there's no way for them to know. . . . So it feels like we're everywhere.

**XB:** You don't know what Extinction Rebellion is going to do next, what city they're going to shut down. We do know. [*All laugh.*] We don't know what Greenpeace is going to do next, what oil-extraction center they're going to shut down next. It's just amazing.

**KE:** The narrative around the climate movement has been "what are we fighting against," and I think that people don't always understand what is that vision that we're fighting for, what is it that we're protecting. And I think that [it's important that we're] painting that picture for people and really talking about climate not only as atmospheric levels of carbon but also as climate justice, and saying we have an opportunity to use the climate crisis as an opportunity to really repair a lot of the broken pieces. And to talk about equity and justice, and to center that in the solutions that we're talking about.

Bringing that into the conversation can often change it from a negative tone into a positive one because people are saying, "Okay, this isn't just about the kind of white, elitist environmentalism that has always existed, but it's about centering frontline communities,

communities of color, communities that are going to be hit first and worst, and ensuring that vision of what could be for them and how they're going to lead the way."

**XB:** Beyond that, Indigenous peoples' voices are being highlighted so powerfully now. We opened and closed our climate strike with Indigenous voices. And they are the protectors of the land. And if we want the land to be protected, we have to protect them. We need to have a whole cultural shift in which it's our culture to take care of the earth—not because it's a movement, but because it's how we were raised. . . . We need to do that through storytelling and show how the climate crisis is affecting real people in real time. When you go to the personal impacts of the climate crisis, or how individual communities are suffering from air or water pollution, and to see that this is about injustice, then people shift from thinking this is about lightbulbs and plastic straws to people's lives. And to shift that narrative, shift the culture is what we're fighting for.

**JM:** I'm super curious how each of you got involved. Have you been involved or aware since you were little?

**KE:** In sixth grade I read [Al Gore's] *An Inconvenient Truth*. I grew up in suburban Wisconsin, and I had never really understood issues to be bigger than a single community. So that was the first time I really understood not just the climate crisis but that there could be an issue facing humanity that was global in scale, and that there was this existential threat.

**XB:** *And* for me, I was born and raised in Mexico, around forty minutes west of Mexico City, in a small town called San Pedro, Tultepec. And my town suffered from heavy rainfall—flooding and contamination that came into the street because factories are pouring waste on the river we have right there. That was the first time I witnessed the climate crisis. My parents raised me being environmentally aware; both of them worked in sustainability development. My dad is Otomi, which is an Indigenous group in Mexico, so their relationship with the land, their relationship with the sacred elements is very strong. And when you see those things being disrespected systematically by society, then you know something is wrong.

And when I moved to New York City, Hurricane Sandy had kind of destroyed many of the seashores in Long Island, and that was the tipping point [for me]—of this is global. This is happening everywhere. The climate crisis follows you. And how can you not do something when this is happening globally? It's also happening to so many communities in so many different ways. The question is, "How can you not?"

**JM:** It's so funny; when I was in college I was studying political science, and I remember somebody saying, "How come you're so involved?" And I remember saying, "How can you not be?" The more you learn and the more you understand, the more, I agree, you have to get involved.

**XB:** It's weird when people say, "Why are you so passionate about the climate crisis?" And I'm like, "This is a ticking time bomb; how are you not trying to dismantle it?"

**KE:** I'm curious to know, from getting to watch and being involved in the climate movement for so long, how you see it shifting and if there are things you're really excited to see or things you wish you could see?

**JM:** I think now what you're starting to see is much more of the connecting and much more of the understanding that we all actually have very similar opponents. The root causes of what we're all trying to address are the same: this kind of short-term, profit-driven economy. In the past I think that was almost too scary to name for some, and I think the movement has gone from trying to defeat coal-powered power plants, plant by plant, to trying to address the underlying root causes.

I also think it's gone from being a very faraway problem to a very current problem. So I think the thing that has an impact on the movement, but also on the whole debate, is what's happening now; is what the scientists thought was going to be happening in ten, twenty, forty years. So I think that's petrifying, but also really makes it very local.

**KE:** I've definitely noticed that. I think for a long time, climate change felt very abstract—this kind of scientific, very futuristic thing.

**XB:** Like big waves on a globe.

**KE:** Yeah, but I think in the last few years, or just months in the U.S. alone, there have been wildfires, hurricanes, flooding in the Midwest. I think they add up. I think people look at their communities and they say, "Okay, all of us are going to be affected in one way or another." So I think it is becoming more concrete in people's minds. We just have to make the solutions more concrete in people's minds, too.

**XB:** We don't want people to have to experience the climate crisis to realize that it is a crisis. And that's what we're fighting for. For less people to be affected and more people to be aware. We want consciousness, we want action, and we want just a complete breakthrough of solutions.

# SECTION 3:
# INTERSECTIONALITY

# Intersectionality in Climate Reporting and Activism

## SAMHITA MUKHOPADHYAY

*May 2020*

*Those least responsible for climate change are worst affected by it.*
**—Vandana Shiva, *Soil, Not Oil***

In 2014, after a series of dubious decisions by Michigan's state leadership, the city of Flint had its water source changed from treated water flowing from Detroit to water sourced from the Flint River. The switch was made without the normal environmental checks: officials failed to ensure the water was even drinkable. It turned out that the water in the Flint River had high levels of chlorides, which destroyed lead pipes; that, in turn, contaminated water being piped into the homes of hundreds of thousands of residents in Flint—a town that also happens to be a predominantly Black and poor white working-class town.

Thanks to local activists, the story of Flint's water crisis made it into the national spotlight and became a seminal public conversation about the role race and class play in environmental issues. And while far more people today may be familiar with the concept of environmental racism, if you ask the average person, they still may not know that there was a town that was deprived of clean water in the United States for years, or the way politics played into how it happened.

Environmental justice activists have long decried environmental racism or the idea that low-income communities of color are more vulnerable to poor living conditions like toxic water, shoddy development, polluted air. Looking at just a handful of other recent cases where the environment posed a serious threat to the communities that lived in them, a pattern emerges. Sometimes, these disasters are man-made, like the Dakota Access Pipeline threatening Indigenous water supplies or a coal plant demolition in Chicago's Little Village neighborhood polluting the air with clouds of dust. Sometimes, natural disasters are exacerbated by human politics, like what happened after Hurricane Katrina in Black neighborhoods in New Orleans or the way Hurricane Maria's devastation of Puerto Rico was exacerbated by neglect. In each case, low-income communities of color were effectively treated as disposable in the environmental economy.

Advocates have developed their own strategies to fight these disparate effects, tactics that recognize that justice alone isn't enough; what is needed is an intersectional strategy to address the fully deleterious effects of climate change and environmental degradation. At the root of it is intersectionality, a term coined by Kimberlé Crenshaw that looks at the many ways power intersects with identity to create social, political, and economic conditions. It is an intersectional lens that allows us to see the spread of Zika virus as a function of both warming temperatures and reproductive rights, that makes sense of the fact that women are often the hardest hit during a climate disaster, too, and that fully realizes that the carnage from Hurricane Maria in Puerto Rico was not just about a natural disaster, but also class, race, and colonization.

Environmentalism has long been focused on protecting the earth: stopping pollution, recycling initiatives, park restoration, and keeping streets clean. Only in the last few decades has there been a shift in the public conversation to issues of intersectionality and justice. Environmental justice has shifted the gaze from a focus on beautification and individual action to a recognition that the climate crisis is due to greedy corporations ignoring warnings about their impacts on the earth. An intersectional lens makes it easy to see that people living in poverty—itself a racialized phenomenon—have no choice but to buy food in wasteful plastic packaging. The most rudimentary class lens reveals that climate change is a real problem, but the communities with the highest carbon footprints are rarely the ones that are on the front lines of the crisis.

The essays in this section apply an intersectional framework to dissect and interrogate the role that power plays in both climate change and environmental activism. These selections explain the basics of environmental racism, document the groundbreaking work of activists of color, report from the front lines on how climate change has impacted Indigenous communities, and look at how those same communities have proposed viable solutions. They detail the brutal realities of the crisis for climate refugees, often women, in the Global South. They explain why the Green New Deal matters and how it connects climate change policy to immigration. They even grapple with how the labor movement must contend with climate.

I would remiss to not mention that while these pieces predated the COVID-19 pandemic, an intersectional lens could have predicted who would be most hit by the public health nightmare: poor communities, overwhelmingly Black and Latino communities, health care workers on the front lines, and essential workers. The coronavirus pandemic is also an environmental issue and a side effect of globalization, and many of the solutions to the hardships people have experienced due to the pandemic lie in an intersectional environmental justice framework.

In the tradition of the famed ecofeminist Vandana Shiva, these essays make the case for how our neglect of the earth is rooted in our neglect of our most vulnerable. Let's hear their voices.

# People of Color Deserve Credit for Their Work to Save the Environment

*December 27, 2018*

We frequently hear that "climate change is everyone's problem." Unfortunately, though, it's a bigger problem for some groups than others in the United States: those most likely to be affected by environmental pollutants are people of color. This phenomenon—wherein environmental risks "are allocated disproportionately along the lines of race, often without the input of the affected communities of color," as the *Atlantic* put it—is called environmental racism.

Environmental racism means that people of color face increased risk of exposure to biohazards, water contamination, and poisons, and they are more likely to be in closer proximity to hazardous waste sites and landfills. These issues arise in communities across the country, in large and small towns, and through a host of mechanisms like polluted water, chemical dumping, and hazardous waste. But even though people of color and low-income Americans are more likely to face these exposures, they are underrepresented in important organizations and mainstream conversations about climate change and environmental protection. This disappointing reality persists at the highest levels of environmental organizational leadership, even in groups that often claim to be inclusive.

The overwhelmingly white, well-to-do caricature of the environmental movement doesn't actually map onto reality. A recent study in the Proceedings of the National Academy of Sciences of the United States of America found major disparities in how much the public believes minorities and low-income Americans care about the environment. Contrary to what most people polled thought, nonwhite groups reported more concern about the environment than whites who were polled. In other data from the Pew Research Center, Hispanic and Black Americans were found to be more likely than white Americans to blame humans for global warming (an ongoing debate within political and climate change circles alike). When studying differences in environmental concerns between Black and white Americans, researchers from the University of Michigan found that the disproportionate burden of environmental pollutants and catastrophes that Black Americans experienced in their personal lives and communities shaped their responses to environmental issues, making them more likely to express concern than whites. So while public perception might suggest that minorities and low-income Americans aren't thinking about the environment, data certainly proves otherwise.

There are many working on the front lines to change the common misconception that people of color aren't environmentally aware.

Jamie Margolin is a teen climate change activist and the founder of Zero Hour, a movement that seeks to "center the voices of diverse youth in the conversation around climate and environmental justice." In an interview with *Teen Vogue*, she explains that the gap between people's perception of marginalized groups' knowledge about the environment and the reality is about "Eurocentric" colonial history.

"People often assume that marginalized communities don't know what we're talking about, but it's actually not true at all," Jamie says. "[They are] implying that people of marginalized identities can't think of anything beyond their marginalized identities."

Beyond the motivation to avoid health risks, people of color and low-income Americans are driven by social and political issues to be attentive to environmental concerns. Environmental justice has become an even greater concern since the election of President Donald Trump, whose policies and agendas on climate change and environmental protection unduly affect Black Americans and other minorities.

At present, the Bayou Bridge Pipeline threatens the lives and livelihoods of residents of Louisiana's Atchafalaya Basin, a water source that provides drinking water, food, and a location for local tourism activities. Young people of color, Native women especially, are defending this space through organized actions, putting themselves directly in harm's way to ensure the safety of the community there. In North Dakota, the Standing Rock Sioux water protectors have been struggling against the U.S. Department of Transportation and other governmental agencies since 2016, trying to prevent the construction of the Dakota Access Pipeline, which Trump "resurrected" after he took office in 2017. For more than a year now, the pipeline has been active on sacred land that the Sioux contend "they never agreed to give up." The water protectors have spent years advocating not only for themselves and their land, but for all humans and all land, reminding us that ending these man-made environmental disruptors and crises is crucial to our survival, because "water is life." These are the voices and experiences that should be central in the environmental movement.

When it comes to plastics and the pollution caused by its creation and waste, the conversation gets complicated. An estimated nineteen billion pounds of plastics end up in the ocean each year—but plastic never fully goes away. Instead, it fragments into smaller pieces called microplastics, which can end up in the bellies of animals and throughout oceanic ecosystems. When consuming fish and salt, humans eat these microplastics, and the full impact on our health is still unknown.

The United States, like other wealthy countries, ships a lot of its plastic waste to other, often poorer, countries. Until January 1, 2018, more than half of the waste that was designated for recycling from the U.S. ended up in China. But now, tighter regulations and restrictions on foreign garbage limit the amounts and types of waste China will accept from its wealthy peers.

This global crisis does not discriminate in its reach, though people of color are still not visible at the forefront of the movement to combat plastics—but that's not for lack of advocacy or care. A 2016 study of California residents found that even though communities of color are more likely to use plastics in their daily lives, they were supportive of taking personal action to reduce the growth of single-use plastics in their communities. After providing participants with more information about how to manage plastics, support for plastic bans increased. These findings suggest that communities of color and low-income Americans, whose recognition within the environmental movement has been almost nonexistent historically, might exhibit more pro-environmental behaviors—behaviors that seek to prevent and reduce the negative impacts of one's own actions on the environment, if provided with greater information and context.

We must consider the marginalized and champion their voices and perspectives on these issues. Take, for example, trendy solutions to combat plastic use, like encouraging paper straws, which recently started to dominate the conversation. Mainstream environmental groups supported the ban, but disability-rights activists quickly highlighted how the focus on banning single-use straws further marginalizes those with mobility and sensory issues. For these populations, the conversation on plastics, and how it factors into larger accessibility issues, hasn't taken full stock of their daily experiences. So any solution, even a popular ban that reduces plastic use, will be incomplete and may reproduce harms against disabled people.

Each diverse population has specific knowledge and expertise when it comes to how best to address the environmental issues facing its community. Until everyone is included in the conversation about the environment, any solutions we land on will fail to fully account for the ways we all move through the world. Not only that, they will reproduce the exclusions and harms against marginalized groups that are already embedded.

This generation of climate justice activists may be the one to end this cycle, and this buzzy moment, when the world is getting smart to the impact of the plastics we use, could work in favor of those most at risk.

According to Jamie, young people in her community are starting to think more dynamically about this issue. "At my school . . . people are constantly worried about plastics," she says. "There's been a lot of focus on plastics lately. That is a conversation starter. That is the tip of the iceberg." She's intent on making sure those conversations don't just include but embrace those who've been left out so far.

But it isn't just about straws, she says. It's about all excess plastic. It's the little bags we use to carry our bananas, a fruit, she points out, that has a "natural wrapper." Even the plastic bags in which we carry snacks to school are more harmful than some people realize, and the impact of their widespread daily use has far-reaching implications and outcomes over time, especially for communities of color. Real solutions to the plastics crisis require we go beyond individual actions, she says.

"There's also this impression, like, 'It's OK if I recycle,'" Jamie says. "Recycling takes fossil fuels. It uses energy." In addition to individual acts like recycling, corporations should be held accountable for their plastic production, which often takes place in low-income areas at the expense of people living there. We should push corporations to find innovative alternatives to plastic production to avoid worsening an already dire situation.

We owe it to ourselves to do better. If you care about the impact of plastic on our natural world and human health, start paying attention to environmental racism and fight back against it. Join causes led by those most impacted, and give them your support as they lead the charge in this struggle for climate justice. Going green shouldn't mean going white.

# What You Should Know about Environmental Racism

LINCOLN ANTHONY BLADES

*December 21, 2016*

Amidst the glow of the holiday season, most Americans find themselves engulfed by the familiarity of joyful tradition and routine, whether it's flying home to see family, or preparing a large feast for ugly-sweater-clad friends. But, in Flint, Michigan, a predominantly Black, predominantly poor city where residents are living with lead-poisoned water, many families are struggling with the tragic reality of newly formed customs such as traveling to water distribution sites just to get clean water, and taking quick showers and praying that the resulting "white blotches" on your skin aren't an indication of kidney failure.

After surviving these horrific conditions for more than two years, on Tuesday, December 20, Michigan prosecutors charged four former government officials including two of the city's emergency managers, Gerald Ambrose and Darnell Earley, in a third round of prosecutions.

Michigan attorney general Bill Schuette charged the former emergency managers with multiple felonies that could result in each of them facing up to forty-six years in prison. Gov. Rick Snyder has yet to be formally charged. Flint's mayor, Karen Weaver, is hoping that this is the start of an investigation that will identify all guilty parties and address the problem of "profit being put over the health and well-being of the people."

The poisoning of Flint's water wasn't an accident or an honest mistake. It was the result of greed and inexplicable shortcuts. It's also important to understand that Flint is actually far from being an isolated incident or even being the worst example of lead poisoning in the nation. A Reuters investigation revealed that there are almost three thousand cities with higher poisoning rates than Flint, most of which are impoverished. Other statistical analysis into the specific neighborhoods and demographics of who've been poisoned reveals that minorities and the impoverished (and especially those lying at the crux of that intersection) are the ones who disproportionately suffer through these problems. This is the face of environmental inequality, and if our society is serious about addressing these issues and their root causes, we must begin by addressing the need for environmental justice.

### What exactly is environmental justice?

Environmental justice is a movement aimed at addressing and abolishing environmental racism and environmental classism.

Environmental racism is the disproportionate impact of environmental hazards on people of color. This occurs when corporations or local, state, and federal governments target and unfairly subject minority communities to unhealthy living conditions. Environmental classism is the disproportionate impact of environmental hazards on low-income people and neighborhoods. This occurs when poor neighborhoods, towns, and cities are unjustly subjected to hazardous surroundings in a manner that wealthier areas aren't.

The Environmental Protection Agency defines environmental justice as "the fair treatment and meaningful involvement of all people regardless of race, color, national origin, or income with respect to the development, implementation, and enforcement of environmental laws, regulations, and policies." Some activists take issue with that definition because they see it as a mandate to poison people equally, while their mission is to ensure that people are not poisoned at all.

## Where did this term come from?

In 1979, the Northeast Community Action Group (NECAG), a group of Black suburban homeowners in a middle-class enclave in Houston, came together to prevent the city from building a landfill near their neighborhood. The group launched a civil rights suit, *Bean v. Southwestern Waste Management, Inc.*, under the legal direction of Linda McKeever Bullard. A report produced in 1979 in support of the lawsuit found that for decades, Houston had built over 80 percent of its landfills and incinerators in predominantly Black neighborhoods. Bullard's husband, Dr. Robert Bullard, began documenting eco-racism cases throughout the city, then throughout the South, and eventually throughout the nation. Their collective actions became a breakthrough moment for fighting environmental decisions as violations of civil rights.

## What are different ways environmental discrimination can be carried out?

Besides placing landfills close to neighborhoods, other examples include placing hazardous waste sites (sewage treatment plants, garbage transfer stations, smelters, etc.) too near to residential areas, and placing other locally unwanted land uses in communities without consulting the residents. Other forms include exposing farmworkers to toxic pesticides, using Native reservations for waste disposal, and denying underserved low-income areas health and safety upgrades such as removing harmful toxins from the air they breathe every day by removing the lead paint in poor apartment buildings.

**What are some examples of environmental discrimination?**
Besides Flint, probably the most famous, recent case of environmental racism is the #NoDAPL protests of the Dakota Access Pipeline. The oil pipeline, which could be disastrous in the case of a spill, was directed away from the predominantly white, middle-class city of Bismarck, and redirected way too close to a Native community, without any requisition or warning. When a corporation decides to make a unilateral decision that threatens minorities' lives without presenting them any options (the same ones granted to white, middle-class citizens), the inherent racial discrimination in their plans is laid bare.

As Dr. Bullard points out, inequities sometimes occur as a matter of class, and thus may be directly targeted at white neighborhoods. "Now all of the issues of environmental racism and environmental justice don't just deal with people of color. We are just as much concerned with inequities in Appalachia, for example, where the whites are basically dumped on because of lack of economic and political clout and lack of having a voice to say 'no,'" he stated in an *Earth First!* interview.

## Why is environmental justice going to be such an important topic in 2017?

Because it appears America's new president is ushering in an administration likely to inflict environment injustice. Donald Trump's cabinet appointments include some who favor the wealthy, some who favor white nationalists, and even an EPA head who tried to sue the EPA—twice. If the policy ideas of Trump's administration are consistent with the cabinet members' history, many of America's most vulnerable citizens will be exposed to harmful threats against their health and safety. The more people understand that racism, classism, and environmentalism are not completely separate and unrelated topics, the greater chance we have at identifying decisions meant to intentionally disenfranchise those relegated to that precarious intersection.

# Brooklyn's Frontline Climate Strike Was Led by the Communities Hit Hardest by Climate Crisis

*September 23, 2019*

The September 20, 2019, youth-led global climate strike brought millions into the streets. Among those demanding action on the climate crisis were youth of color from around the world, who gathered in the DUMBO neighborhood of Brooklyn to spearhead their own fight: the Frontline Climate Strike to create an intentional space for the communities of color often hit the hardest by climate disaster as part of the Climate Justice Youth Summit.

With chants such as "What do we want? Climate justice! When do we want it? Now!" live music, and lit sage, intergenerational strikers marched to Brooklyn Bridge Park, where youth of color from places like Michigan, Indiana, California, and more shared stories of what they are fighting for.

158

Each climate striker built on the story of the last, articulating experiences using terms like "just transition" and "environmental racism" and expressing their desire to move from an "extractive economy" to a "regenerative" one. Overall, the young people denounced "false" uniform solutions in favor of a community-tailored approach. They powerfully drew on their experiences at the intersections of race, class, and climate vulnerability to advocate for a future where those factors will be considered in any crisis response.

## What Are Frontline Youth Fighting For?

Nyiesha Mallet, at the time an eighteen-year-old artist and activist from the Brownsville neighborhood of Brooklyn, boils down the frontline youth fight to this axiom: climate justice entails an economic shift that swaps an economy focused on the individual for one focused on people. Beyond clean energy, that involves a just transition, which means helping phase out environmentally harmful industrial practices for better pathways, while also making sure workers from those industries aren't left out in the cold. And she believes people of color know just how to do it.

"We need people to start listening to frontline community members because we are the ones with the solutions, not the people higher up, who're looking out on the map and judging us based on [our experiences, saying,] 'There's a heat wave here, and there's water melting here,'" she tells *Teen Vogue*.

"Young people of color have been doing the work to fight climate change for hundreds of years. It's something that we are born into," she says. "We've lived in an extractive economy our entire lives; we come from a generation of families that have to live through this extraction and we know what it is, we know how it affects us, and we know what kind of change we want to see."

## More Than Local: Honoring Intersectionality

For Chelsea Turner, an artist and community youth organizer of Barbadian and African American descent who lives in New York, her hope is that frontline youth understand that their lower-income communities of color are in the places where climate disasters can be devastating long-term, even causing mass migration.

"That means that they will be hit first and worst, and are least likely to get government assistance in the event of extreme weather events," she says. "And they're also, really, now, being affected by active environmental racism for generations. In our communities we have a lack of green space, lack of access to healthy foods, or affordable healthy foods, which are known as food deserts. We have a lot of highways, which increases asthma rates, as well as nuclear power plants, and a lot of power plants in general in these areas, and we are fighting back against them."

The young activists were joined by other youth representatives from Guam, Mexico, Thailand, and Honduras, there on behalf of an organization or independently. Several had stories to share, and though their details differed, they all described similar fights against the same cultural forces. Areerat "Aree" Worawongwasu, a young activist from Thailand, who attended the strike with her mother, honored her ancestors and shared a moving account of her fight alongside other frontline community members.

"A lot of waste from the Global North is dumped here," she says of her home city, Bangkok. "Our livelihoods are tied with climate.

This isn't just a hot issue for us, this isn't just about statistics; this is about our livelihoods, and we're fighting at the front lines."

The impact of climate change has also managed to sound an ominous alert for the U.S. Indigenous community, and prompted Simon Montelongo, a Cherokee person from the Qualla boundary in North Carolina, to join the Frontline Climate Strike in Brooklyn. For Simon (who uses they/them pronouns), uniting forces is about more than solidarity; it's also a form of self-preservation.

"With all of these recent fires, oil refineries, and all of these recent fracking ideas, our communities back home are being pressured, because we are one of the last open, free natural resources in the area," they tell *Teen Vogue*. "If they continue to use and degrade and disregard all of their resources, then they're going to be looking at our communities next."

*Star Wars* actor Oscar Isaac, who has played rebel leader Poe Dameron in the latest trilogy of the film series, made a notable cameo representing Guatemala and thanking the youth for taking a stand for climate justice. "I want to honor you guys and say I'm so proud—as a brother of color, as well—to see all of these young people out here fighting for climate justice," he said.

## Puerto Rico, Presente

A major part of this cross-cultural and intergenerational stand against the climate crisis was honoring the second anniversary of Hurricane Maria in Puerto Rico. Protestors observed a moment of silence to honor the victims and gave their platform to those who hailed from Puerto Rico.

Luis Collazo Colón, a seventeen-year-old from San Lorenzo, Puerto Rico, took the stage to thank the Climate Justice Youth

Summit for the hard work and to entrust other teens present with helping him build a better world. Aside from the repercussions of limited aid during and after Hurricane Maria, Colón took the opportunity to highlight his town's most urgent struggle: a defunct waste-disposal system. San Lorenzo, a land that's split between three municipalities, is one of the towns that's struggling with abandoned sick and dead animals and random trash disposed of in front of residential homes.

"I hope that the government we elect can bring change to my community," Colón tells *Teen Vogue* at the September 20 strike. He shares how he's personally seeing people who are still trying to cope with the storm's aftermath, saying, "As of now, my mother and I nurture animals such as sick horses, or whatever passersby dump. But maybe animal-rescue organizations can swoop in and help with relocating abandoned and stray pets."

## United We Stand

These youth of color are not alone in their fight. Whether outside the movement or inside, adults are playing their role in supporting as they deem fit. "It wasn't until college that I was exposed to a lot of the realities of people outside of the whitetopia, that my family, as Italian immigrants, chose to move into," Frank Marino, a Brooklyn-based humanities teacher at MS 839, tells *Teen Vogue*.

Therefore, he chose to contribute by exposing eighty middle school students to the frontline strike organized by a local grassroots organization. He hopes to educate his "privileged" white and wealthy students through this exposure, and to show youth of color that there are people who look like them fighting for climate change.

Meanwhile, long-term environmental activists of color have been mentoring the youth to ensure that the new generation doesn't experience pitfalls. Kari Fulton is one of the adults who encouraged the youth to call upon their ancestors, insisted on the use of proper pronouns and respect for cultural diversity and kept the chants alive during the march. "Our job is to make the space for folk to walk in and have the notes and know where the land mines are, so they can navigate," she tells *Teen Vogue*.

# How Climate Change in Bangladesh Impacts Women and Girls

### KAREEDA KABIR

*July 16, 2018*

When people think about the impact of climate change, many consider the physical damage: homes destroyed, communities forced to start over, maybe even a number of bodies discovered after an intense weather event. But sometimes forgotten are the social consequences the physical destruction leaves in its wake.

In Bangladesh, one of the main social challenges presented by climate change is the furthered entrenchment of preexisting systemic gender inequality. As climate change negatively impacts vital local industries such as rice farming and fishing, women and girls experience a compound set of issues.

Flooding, a result of higher aerial moisture levels combined with increased runoff from the Himalayas, can lead to disaster—especially in a country like Bangladesh, where nearly half the population lives, on average, just ten meters above sea level. In June, five Bangladeshis were killed by massive flooding and more than fifty thousand Bangladeshis have been affected. Flooding, often a product of climate change, can deepen gendered social problems, too.

Flooding often leads to salinization, or the increase of salt levels in freshwater sources. Once water has been salinized, it can't be used for drinking or farming. Water-related health complications are already a major issue in Bangladesh, and the lack of usable, clean water exacerbates this. These issues are likely even worse for women. According to a 2016 study published in the *Journal of Environmental Protection*, "Higher rates of pre-eclampsia and gestational hypertension in pregnant women in coastal Bangladesh, compared with noncoastal pregnant women, were hypothesized to be caused by saline contamination of drinking water."

Flooding can also cause the areas around bodies of water to erode, which often leads to siltation, an environmental process in which excess minerals enter water. As the *Daily Star* reported in 2016, siltation has significantly contributed to the disappearance of thirty-two fish species from Hakaluki Haor, the largest body of water in Bangladesh. The multifaceted impact of environmental deterioration on women in the country can be seen here: as pointed out by Melody Braun of the nonprofit research organization WorldFish, dwindling fish populations not only drive vitamin and mineral deficiencies in women (who often prioritize the nourishment of their husbands and children), this problem also hampers the economically essential practice of fishing. For women who work in Bangladesh, as fish disappear, so do their job opportunities. Relying on fish to sustain themselves dietarily and monetarily, Bangladeshi women and girls experience acute social costs as a result of decreases in local fish populations, caused by problems associated with climate change.

The negative and gendered implications of climate change in Bangladesh are not solely demonstrated through health-related consequences; women and girls endure a steep social cost as the climate deteriorates. Displaced by massive flooding, most villagers migrate toward cities (many to Dhaka, the nation's capital) in search of work, where secondary school can be prohibitively expensive.

If Bangladeshis choose to stay in their villages and rebuild, families will often start looking to get their children married. With the average age of marriage for a Bangladeshi girl hovering around sixteen years old, according to Plan International Bangladesh, marriage exists as a major limiting factor where the education of girls and women is concerned. Nevertheless, the prospect of marriage is seen by many as a win-win situation for both families: the groom's family gets a large sum of money or gifts (traditionally, the bride's family pays a dowry to the groom's), which can be used for home repair, preparing for more flooding, or other living expenses; and the bride's family, to put it crudely, has one less mouth to feed.

Worse still, where flooding is rampant, schools are often the first public spaces to be transformed into makeshift shelters where displaced villagers can stay as they work to rebuild their homes. This pushes rural children, especially girls, for whom education is already barely within their grasp, away from schooling. Homes can be rebuilt, and temporary jobs can be found; however, being denied an education at crucial periods in these women's lives can have lasting, irreversible impacts.

While it is essential to take every step possible to prevent natural disasters like the massive flooding we've come to associate with climate change, it is equally essential to deal with the social consequences environmental disasters create. Organizations like BRAC Centre have shown this work is not only critical, but also fully realizable, managing to enroll two million girls in school in 2017 alone. But BRAC cannot do this work alone. In the global fight for the recognition of women's issues, climate change as both cause and catalyst of furthered subjugation cannot be overlooked.

# Climate Change Is Creating a New Atmosphere of Gender Inequality for Women in Malawi

### MÉLISSA GODIN

*December 20, 2018*

"I used to live here," says Sofia as she stands on a raised mound of dirt still imprinted by the bricks that used to be there. "But my home was swept away during the floods in 2015. I lost everything."

Sofia is a Malawian woman in her thirties living in a remote village in the southern district of Chikwawa. Like many Malawians living in the Lower Shire Valley in southern Malawi, Sofia's life has been severely affected by climate change.

Most Malawians, however, have done little to contribute to global warming: Malawi ranked 162nd for carbon dioxide emissions as of 2010, making it one of the least significant polluters in the world. According to the World Bank, only 11 percent of Malawians have access to electricity. Nevertheless, the United States Agency for

International Development (USAID) considers the country's population as especially affected by and susceptible to climate change, due to its dependence on agriculture—80 percent of Malawians work as farmers with small operations. *Business Insider* reports that Malawi is one of the poorest countries in the world in terms of GDP per capita, depending heavily on agriculture. Household income is thus severely affected by erratic environmental patterns that affect crops, leaving families with little money and little to eat.

"Because of changes in the environment, it is as if we are moving backwards in our development," explains one woman.

The impacts of climate change, however, are not felt equally by all Malawians. Women, in particular, are facing drastic changes to their lives because of environmental changes. For instance, it has been reported by the *Guardian* that the effects of climate change increase the frequency of early marriages. While men are also affected by climate change unseen in many parts of the developed world, it is often easier for men to adapt to climate change due to their greater freedom of mobility and fewer household responsibilities.

In many parts of Malawi, women are often responsible for fetching water and firewood. Due to the current drought and massive deforestation throughout the country, women have to walk longer distances to access these basic necessities. This not only increases the amount of time women spend on unpaid labor but also puts them at risk of sexual violence.

Malawi, which used to be covered in forest, is facing massive deforestation throughout the country due to excessive logging as well as "slash and burn" agriculture, a method for growing food that involves wild vegetation being cut and burned.

"We walk longer distances for firewood and water. We now walk for four hours," explains Sofia. "And sometimes, when we reach the forest to fetch for firewood, we get raped." She says that pregnancy and HIV infection are possible long-term risks of these attacks.

Because Sofia now spends the majority of her day fetching firewood and water, she has less time for income-generating activities. While she used to spend her days selling firewood and food in the market, Sofia now struggles to ensure that she has enough food to feed her family.

"I do send my kids to school but sometimes when there is no food at home, I tell them to come with me to collect food or firewood," Sofia remarks.

Teenage girls have been particularly affected by the drought. With 46.34 percent of the Malawian population under the age of fourteen, with a median age of 16.7 years for women, teenage girls constitute a large segment of the Malawian female population. Teenage girls, however, face unique risks compared to older women. During times of scarcity, young girls are expected to assist their mothers with household chores, often resulting in girls dropping out of school. Although education levels are low in rural parts of Malawi for both men and women, AGE Africa notes that girls are less likely to be in school than boys. Due to climate change–induced poverty, many young girls have decided or been forced to get married as a way of coping with increased financial vulnerability.

This was the case for Maria, a woman in her early twenties living in a village in the district of Chikwawa.

"At my home there were no basic necessities, so I got married at fifteen to provide for myself," explains Maria. "My parents were farmers and their seeds kept drying up or getting washed away during floods."

Although Maria is now happy in her marriage—and has access to food and better shelter than she did at home—she wishes she could have completed her education.

"I got married because of climate change," she explains.

Maria, however, is not alone.

"Climate change has caused women and girls to enter into early marriages because parents are unable to support their families," she says.

Teenage girls are also more likely to face challenges during times of environmental disasters compared to older women or their male counterparts.

PLAN International reports that during environmental disasters, adolescent girls are particularly at risk, often facing sexual violence, unwanted pregnancies, and school dropouts. With environmental disasters likely to become more frequent and severe in Malawi, young girls will face more challenges.

"When we don't have floods we have drought," explains Sofia. "And when we don't have droughts, we have floods."

Before the floods in 2015, Sofia lived by the Shire River, where she owned her home, had access to fertile land, and could feed her family. Though floods were a regular occurrence, they tended to be smaller and less frequent. Some of the women interviewed said that communities living along the water relied on Indigenous knowledge systems that helped them anticipate and prepare for floods. Many women remarked that these techniques are no longer appropriate in this new climate of violent environmental disasters.

After her home was swept away, Sofia had to move upland, where she now rents a thatched-roof home. In this new community, Sofia has no access to fertile land to farm.

"Now, the breadwinner is my husband," Sofia explains. "Women are now voiceless because men have access to money, they have access to everything. So they take advantage of that. When we are out collecting firewood, they stay at home and do nothing."

Sofia's husband recently married a third wife. When there is no food in Sofia's home, her husband stays with one of his other families because they can feed him. Many men, however, believe they are equally or more affected by climate change than women.

"The environment affects me in a special way," explains Maria's husband. "As the head of the household, I am responsible for everything—I have to make money. My wife and family depend on me. Because we can no longer depend on farming, it is up to me to find money, to bring food to my wife and kids. So I find this very difficult."

Though men are also severely affected by climate change, they tend to have greater opportunities to adapt to environmental changes than women. Many men are aware of these gendered dynamics, noting in interviews that women's household responsibilities have become more burdensome due to environmental changes. While women have to stay near the home to provide for the family, men can migrate to other areas in search of work or greener pastures, sometimes leaving behind wives whose land is confiscated because, in some patrilineal districts in Malawi, it is perceived to have belonged to the men.

Male migration has subsequently become a source of vulnerability for women, with many men leaving and never coming back.

This leaves women financially vulnerable and susceptible to stigmatization as single women within their communities. Though some women speculated to *Teen Vogue* that their husbands have remarried and forgotten about them, others think that their husbands have been unsuccessful in their quest for work and have not come home due to lack of funds or embarrassment about their inability to provide for the family.

For Sofia, there are tangible ways of mitigating the impacts of environmental changes.

"We need irrigation farming," one of many women interviewed by *Teen Vogue* said, expressing a common sentiment several women shared. "We need knowledge and technology."

Despite there being a huge presence of NGOs throughout Malawi, many communities feel as though the response to climate change has been slow and reactive. While the country has implemented climate change policies, such as criminalizing the illegal cutting down of firewood and charcoal-based livelihoods, these policies have hurt the poor who need firewood to survive. Rather than providing communities with alternative strategies, many communities feel as though climate change policies further stigmatize them. Similarly, several experts that spoke to *Teen Vogue* expressed frustration about the lack of attention paid to climate change's gendered implications.

Though the government and aid organizations have responded to environmental crises by providing food aid and basic shelter, many women told *Teen Vogue* that they feel little has been done to ensure that communities and women specifically have the skills and resources to adapt to climate change. Research for this article found that out of twelve villages visited in southern Malawi, only two had received assistance in developing adaptability strategies. In one community, an organization had started a solar energy project in 2014 that was never completed. In the other, unoccupied, half-finished "flood-proof" homes littered the landscape while community members continued to live in houses prone to collapse.

Several communities are consequently disenfranchised by aid organizations and the government, based on the assumption that they are corrupt and self-serving.

With the help of other women in her community, Sofia puts on plays and dances that bring the devastation of the flood that swept away her home and the ineffective subsequent response to dramatic forms. She believes that by raising their voices and honoring those that lost their lives to the flood, her community will always remember how environmental change has affected their lives.

"I don't have hope that the future will be like what it was before," reflects Sofia. "If God allows us to live like we did before, then we will. If God does not, I will continue to live this hardship."

# Climate Change Is Impacting Indigenous Peoples Around the World

## MICHAEL CHARLES

*March 5, 2019*

On October 8, 2018, the Intergovernmental Panel on Climate Change released a report on climate change. It showed the consequences of inevitable global warming and the drastic action needed to meet the goals of the Paris Agreement, which aims to limit global warming to 1.5 degrees Celsius (2.7 degrees Fahrenheit) above preindustrial levels. Headlines popped up all over news sites and Facebook feeds, and "twelve years to act" became the topic of conversation. Panic and worry spread through communities over the potential for "future" climate crisis. However, many people don't understand how Indigenous communities are already facing this crisis—right now, today.

Although there is no universal definition for Indigenous, and Indigenous peoples are self-identified, among us we share many common characteristics. For example, most of us have Indigenous languages, land that we connect to and identify with, and stories and cultural practices that are passed down and used in our daily lives. Many of our cultures and languages are deeply tied to the way we know and interact with the natural world around us.

When I first learned the Diné word *kéyah* (land), it was described to me as being the connection between our feet and our Mother Earth. Similarly, when I spoke with an iTaukei, or Fijian, friend, he shared a word from their language: *vanua*. From my understanding, *vanua* is a holistic concept that describes humans, the natural world, culture, and, as a result, identity. I have learned of many similar concepts across Indigenous cultures, and it reveals a major impact that climate change has on Indigenous peoples.

Our Indigenous cultures have taught us through our languages, stories, and practices that our identity is tied to the land. So what happens to us as the climate changes? Our identity is impacted. The ways we know how to farm, fish, and hunt must also change. Our practices, languages, and stories will need to adapt to explain the new relationship or identity we experience through climate change. Most of our Indigenous knowledge systems are a result of our ancestors' collective experience over millennia and teach us how to survive and provide for future generations. What we are experiencing now is a rapid change in climate that is unprecedented, impacting the way our existing knowledge systems can accurately teach us about the world around us.

I am fortunate to have the opportunity to work with the International Peoples' Forum on Climate Change, a community with whom I have been able to learn about how climate change impacts my global

Indigenous family from across our seven self-organized regions: Africa, Asia, the Arctic, the Pacific, North America, Latin America and the Caribbean, and Russia and Eastern Europe.

I mentioned how, recently, widespread panic and worry have occurred, thinking about the 1.5 degrees Celsius mark of global warming. According to the WWF Arctic Programme, the average temperature of the Arctic has increased 2.3 degrees Celsius since the 1970s, impacting the livelihood and safety of approximately four hundred thousand Indigenous peoples who live there, such as the Saami, Inuit, and Nenets. As temperatures warm, the ice thins and causes displacement of the people from their homelands and displacement of species that are fundamental to their diet and economies, and it has increased the number of reported ice accidents resulting in injury and death. For Indigenous peoples in the Arctic, climate change is directly impacting them today—their safety, economies, cultural practices, and Indigenous knowledge and identity.

But this story does not stop here. Remember, everything is connected.

As the ice melts, sea levels rise and impact Indigenous peoples who live on islands or coasts. One of my Maori friends from New Zealand has told me that their home will be underwater in their lifetime. Many people along coastlines have to adapt as sea levels rise and move further inland.

However, to many Indigenous peoples, it's not just land that is submerging, it's part of their identity. While some Indigenous peoples will worry about managing water, others will worry about managing without it. Increased drought is impacting my people across the Navajo Nation in the Southwest U.S. As water is becoming more scarce, farming and shepherding are being impacted, along with the survival of wild horses throughout our land. Similar issues associated with drought are impacting Indigenous communities throughout Africa.

Economics and politics associated with climate change also heavily impact Indigenous peoples, because, according to the World Bank Group, Indigenous peoples account for 15 percent of the extreme poor, relationships with colonial-nation states often violate our human rights, and treaties meant to recognize our sovereignty are frequently broken. A 2018 paper by the international Indigenous rights organization Cultural Survival reported that China hadn't recognized the term "Indigenous" within its borders; and in Latin America, Indigenous peoples' land rights and safety are at risk from the exploitation of the Amazon. Indigenous activists are fearing for their lives in these areas as the number of murders against Indigenous activists is rising. In 2017, 201 environmental activists were killed, according to a report by the human rights organization Global Rights.

There are many more unnamed examples of the impacts of climate change on Indigenous peoples, and even more examples of Indigenous resistance, resilience, and adaptation. We are often seen as victims in the discussion of climate change, but we have solutions. These solutions are within our cultures, our languages, our songs, our ceremonies, and our spirituality.

We are guardians of this Earth, as the World Bank Group noted in a 2008 paper acknowledging that we protect 80 percent of global biodiversity. Each of us has our own story, and we cannot be understood simply as "Indigenous peoples." Most Indigenous peoples alive today exist because of the survival and resilience of their ancestors through colonization. And we're still here. I want to leave you with the same conclusion that I delivered to global leaders at the United Nations climate change conference, Climate Negotiations, in December:

"The resilience of my people has enabled my existence. Listen to Indigenous peoples' voices and humbly accept our wisdom and knowledge as solutions moving forward, so it can also enable yours."

# The Red Deal Is an Indigenous Climate Plan That Builds on the Green New Deal

RAY LEVY-UYEDA

*November 1, 2019*

As the climate crisis continues and public calls for action mount, the Green New Deal has been centered as a broad proposal for an array of policies that could address the man-made impacts on the climate. But a response has arisen amid concerns that the various programs embodied in different versions of the deal could leave Native climate needs and understanding of the earth out of the conversation. That response is called the Red Deal.

The version of the Green New Deal introduced in February 2019 by Representative Alexandria Ocasio-Cortez (D-NY), which calls for a nationwide approach to climate change to avert a potential climate disaster, drew inspiration from the Green New Deal originally popularized by the Sunrise Movement, a youth-led grassroots organization that built upon organizing knowledge of Native, Black, and brown communities.

The Green New Deal calls for clean-energy jobs, infrastructure, decarbonization, and support of vulnerable frontline communities. Included in the list of frontline communities are Indigenous folks, the land's first people, who only see a dedicated section to the history and destruction of community, culture, and health on the last page of the bill, H. RES. 109, 116th Congress 1st Session.

That section reads: "A Green New Deal will require the following goals and projects . . . [including] obtaining the free, prior, and informed consent of Indigenous peoples for all decisions that affect Indigenous peoples and their traditional territories, honoring all treaties and agreements with Indigenous peoples, and protecting and enforcing the sovereignty and land rights of Indigenous peoples."

For Cheyenne Antonio, that's not good enough. That's why she and other organizers with the grassroots Native organization called the Red Nation have proposed a Red Deal, intended not to replace but to support the proposal from Ocasio-Cortez and others by centering Native leaders and the knowledge that comes with centuries of fighting back against a government that sought to destroy them.

Water protectors and land defenders who live on the front lines of environmental degradation and are often the first to protest its destruction are mentioned once in what is being heralded as the revolutionary policy our country needs right now. So what are they asking for with the Red Deal? *Teen Vogue* spoke with Antonio and Red Nation cofounder Dr. Melanie Yazzie to find out.

## What Is the Red Deal?

The Red Deal was crafted by community members, Native people, young people, and poor people. It has four key tenets designed to build on and push forward the ideas in the Green New Deal: first, what creates crisis cannot solve it; second, change must come from below and move to the left; third, politicians can't do what mass movements do; and fourth, the climate conversation must move from theory to action.

"We draw from Black abolitionist traditions to call for divestment away from the criminalizing, caging, and harming of human beings AND divestment away from the exploitative and extractive violence of fossil fuels," the Red Nation writes of the first tenet on their website. "Proposed discretionary spending for national security in 2020 comes in at $750 billion. . . . And only $66 billion of discretionary funds are spent on healthcare each year. . . . This proves there is an overabundance of energy and resources that go into demonizing Indigenous water protectors and land defenders, Muslims, Black people, Mexicans, women, LGTBQIA2+, and poor people."

In other words, it's not enough for Red Deal proponents that the Green New Deal seeks to create jobs in renewable energy and pushes for access to clean water, food, and a livable planet.

Responding to the fact that the bill does not call to end fossil fuel consumption, Antonio said that the Green New Deal language could create potential for "normalizing fracking again [and] normalizing nuclear again, and doesn't give an option for our people or the planet." She explained that Red Deal proponents believe the Green New Deal should include language that explicitly bans fracking and every form of resource extraction.

The second tenet of the Red Deal, advocating for change from below and to the left, is a call to recognize the inspiration for the Green New Deal and to go further. In order to take the next step toward climate justice, the Red Deal states that a mass movement is needed: "We must throw the full weight of people power behind these demands for a dignified life. People power is the organized force of the masses; a movement to reclaim our humanity and rightful relations with our earth."

The second tenet of the Red Deal, which claims that mass movement politics must be a catalyst for change, underlies the third tenet, which suggests that politicians cannot save our planet by attending to symptoms of the problem. Incremental political reforms do not address underlying causes, and neglect to hold to account industries that perpetuate the climate crisis.

"[Gradual reform] attempts to treat the symptoms of a crisis, rather than the structures of power that create crisis in the first place," the Red Nation writes. "Reformists misunderstand this fundamental truth about capitalist states. States protect capital and the wealthy class, not life. Reformists who appeal to the state for change compromise our future. We refuse to compromise."

"We actually have to build really broad, mass coalitions," Yazzie explained, connecting tenets two and three. Yazzie noted that politics takes note of "what's happening in the public sphere" and added, "The politicians will respond and follow our lead anyway; we're really interested in empowering people to feel like they can own this work."

With its first three tenets laying out that the system is broken and only a mass movement—not piecemeal political reforms—can fix it, the Red Deal's fourth tenet is a call to apply that understanding in action.

"We must reclaim our collective power," the Red Deal's fourth tenet reads. "When the state invests its greatest resources to contain the threat of mass mobilization, we must already be organized in those spaces and those communities. We must be one step ahead, ready to capture the momentum of the next rebellion and catapult it into a full-blown mass movement."

"We don't have time for people to pretend that the problem is much further off," Yazzie said of the fourth tenet. "We have to collapse the process and figure out ways to move forward together." She explained that that the Red Deal is intended to empower people at the local and regional level in order to tackle the "urgent issues that they're facing in their local context."

"We need people as a whole to lead these conversations and move it beyond legislation," Antonio told *Teen Vogue*, stressing that there is no time to wait. "At this point, we need action. At this point, we need people organizing in their communities."

## Why Does It Matter?

The Red Deal asserts that the fight for climate justice must center Native people when it comes to the issues that disproportionately impact Native communities, but it also communicates what the Green New Deal does not—namely, that public lands are stolen lands and climate change is significantly caused by just a few industries, which the government has at worst neglected to hold accountable and at best assisted in their efforts to mine the earth for resources in a move that put profits over people.

"We need to center Indigenous people who've been at the forefront," Antonio told *Teen Vogue*. Native people have the worst health outcomes of any demographic group in the country, and their communities are at times on the front lines of places where government and private companies pollute. Antonio believes that a Native-centered climate justice approach will not only aid the environment but also address the high rates of deadly diseases and cancers brought on by uranium mining and fracking, just to name a few.

"We envision [that] the Red Deal is tying the land and body violence [together]," Antonio said, speaking to the connection between ecological harm and environmental health hazards.

The land violence she's talking about is not just the more than five hundred treaties signed and broken by the United States to build the country by usurping territory, but the attitude and actions of the U.S. Bureau of Land Management (BLM). The BLM has continued to lease the land nearby sacred Chaco Canyon in New Mexico (some of which is a "protected" national park) despite protests by organizations like the Red Nation and Navajo government officials and environmental groups calling for the BLM's findings on environmental assessments to be rejected.

The Red Nation has also protested the use of new extraction technology such as horizontal hydraulic fracturing, which Antonio said the BLM conducted without the consent of Native communities. (*Teen Vogue* has reached out to the BLM for comment.)

While the erasure of Native people and lack of consideration for sacred sites is nothing new to Native communities, the rate at which the BLM is leasing land without Native consent is concerning because once the land is destroyed, there's no getting it back. The Trump administration has opened up a record number of protected public lands for oil and gas exploration and extraction, which has a direct impact on the health outcomes of Native people cultural and historical preservation, and which adds to a history of normalized corruption, genocide, and "ongoing behavior of colonial government trying to erase us," Antonio explained.

Indigenous climate protection actions have been at the center of the Green New Deal presented by AOC. In a *Time* cover story earlier this year Ocasio-Cortez said a trip to Standing Rock was "transformational" on her way to becoming a climate policy champion. Now, Red Deal proponents hope they can push beyond the language included in her version of the Green New Deal for a version that accounts for their concerns.

Antonio told *Teen Vogue* that AOC gives her hope and she admires the congresswoman's strength, but that "she could do better. We can all do better."

### How Does It Move Forward?

The Green New Deal will remain a proposal, barring Democratic takeovers of the Senate and White House in 2020, and even then, there's no guarantee the proposal would become policy or even look the same. But Antonio and Yazzie believe that whatever happens with the Green New Deal, the Red Deal is a first step in centering Native people and offering ways for non-Native people to hold themselves and those around them accountable.

Understand how you're "accountable for the violence that's happening," Yazzie said, or "find out whose land you're living on," Antonio said, explaining, "Wherever you're at, you're on Native territory."

Native people face violence every day, Antonio said, and it will be people power, community organizing, and uplifting marginalized

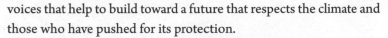

voices that help to build toward a future that respects the climate and those who have pushed for its protection.

"Ours is the oldest class struggle in the Americas; centuries-long resistance for a world in which many worlds fit. Indigenous peoples are best suited to lead this important movement," the Red Deal reads. Or, as Antonio put it, "[Indigenous peoples] have been at the forefront and on the front lines in holding the United States accountable. [We've] been fighting for five hundred years."

# How Plastic Is
# a Function of Colonialism

## DR. MAX LIBOIRON

*August 22, 2019*

Nain is the most northern Inuit community in Nunatsiavut, Canada. It was one of the first places in Newfoundland and Labrador to ban plastic grocery bags in 2009 after villagers saw hundreds of plastic bags snagged on rocks underwater when they went out to fish. The bag ban has appeared to reduce the number of grocery bags in the water, but many of other types of plastic bags, as well as food packaging, ropes, building insulation, and tiny unidentifiable fragments, line the shores and waters of the area.

None of these plastics are created in Nain. But since plastics have been found in the Arctic, government and scientific projects are looking to find ways to reduce plastic pollution coming from Arctic communities with initiatives like recycling and treating sewage. But these solutions look at the end of the pipeline—the point after plastics have already arrived, thousands of miles from their

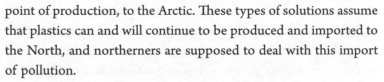

point of production, to the Arctic. These types of solutions assume that plastics can and will continue to be produced and imported to the North, and northerners are supposed to deal with this import of pollution.

"Colonialism" refers to a system of domination that grants a colonizer access to land for the colonizer's goals. This does not always mean property for settlement or water for extraction. It can also mean access to land-based cultural designs and culturally appropriated symbols for fashion. It can mean access to Indigenous land for scientific research. It can also mean using land as a resource, which may generate pollution through pipelines, landfills, and recycling plants.

Lloyd Stouffer, editor of *Modern Packaging* magazine, declared in 1956 that "the future of plastics is in the trash can." This call for the "plastics industry to stop thinking about 'reuse' packages and concentrate on single use" came at the start of a new era of mass consumption of plastics in the form of packaging, which now accounts for the largest category of plastic products produced worldwide. He saw that disposables were a way to create new markets for the fledgling plastics industry.

This idea assumes access to land. It assumes that household waste will be picked up and taken to landfills or recycling plants that allow plastic disposables to go "away." Without this infrastructure and access to land, Indigenous land, there is no disposability.

Nain does not have an "away." Neither do many other places whose lands are colonized as places to ship disposables or are used for landfills. Nor do many extractive zones that provide the oil and gas feedstock for producing plastics. They're in the far North, Southeast Asia, and western Africa, among many other places. Some of these same places serve as an "away" for wealthier regions that export their waste. In fact, the term "waste colonialism" was coined in 1989 at the United Nations Environmental Programme Basel Convention when several African nations articulated concerns about the disposal of hazardous wastes by wealthy countries into their territories.

China has been the place where nearly half the world's plastic waste has been sent to go "away." This ended in January 2018 when China banned the import of scrap plastics and other materials, which will leave an estimated 111 million metric tons of plastic waste displaced. Recycling programs in the United States and around the world that depend on using other countries' land for waste have slowed down, shut down, or are stockpiling plastics as new solutions are sought. Currently, this next round of waste colonization is headed for Southeast Asia.

Perhaps you've heard that the top five countries responsible for most marine plastics are China, Indonesia, the Philippines, Vietnam, and Sri Lanka. Some of these countries are also the ones receiving a disproportionate amount of plastic waste from other regions. They also happen to be places where waste systems do not mimic American curb-to-landfill systems. These regions are framed in scientific articles, the media, and policy papers as "mismanaging" their waste. This is a perpetuation of colonialist mind-set, discourses that have long associated some uses of land as civilized and moral and other uses as savage and deficient. As Cole Harris writes in his book *Making Native Space: Colonialism, Resistance, and Reserves in British Columbia,* historically, when local people were not using land "properly," colonizers would come and take it away to use it "better." In 1876 a white Indian reserve commissioner on Vancouver Island in the region currently known as Canada addressed members of "a Native audience" (nation unspecified) who were being moved to reserves that were a fraction of the size of their previous land bases. He explained, "The Land was of no value to you. The trees were of no value to you. The Coal was of no value to you. The white man came. He improved the land. You can follow his example." Similar mind-sets still exist today.

In September 2015, an American-based environmental NGO called the Ocean Conservancy released a report looking for solutions to marine plastic pollution. One of the core recommendations was for countries in Southeast Asia to work with foreign-funded industries to build incinerators to burn plastic waste. This recommendation follows a long line of colonial acts from various entities, from access to Indigenous land to extract oil and gas to make plastics, to the production of disposable plastics that requires land to store and contain them, to pointing the finger at local and Indigenous peoples for "mismanaging" imported waste, and then gaining access to land to solve their uncivilized approach to waste management.

The Philippines arm of the Global Alliance for Incinerator Alternatives (GAIA), a grassroots environmental-justice coalition, rejected the Ocean Conservancy recommendation for incineration. They argued against the health and environmental impacts of burning waste, particularly in countries that struggle with air pollution, such as China, where increasing protests against waste-to-energy incinerators occur in a context where 69 percent of current incinerators have records of violating environmental air pollution standards. They talked about the costs of building and maintaining this infrastructure and what it means for debt to foreign bodies. They wrote about how burning waste and plastic perpetuates climate-changing fossil fuel extraction. In short, they argued against the entire system that assumes access to land for come-from-away industry and environmentalists. GAIA's efforts have been uneven. They've helped to successfully block some incinerators, such as one that was planned in Wellington, South Africa, and continue to battle on other fronts.

Disposability is not the result of the bad behavior of some individuals choosing to buy some things and not others. Consumer choice as a concept makes no sense in many places. In Nain, there is one store. There is one kind of ketchup you can buy. There is one type of lettuce. Both are in plastic packaging because the producers assume that there is a place for that packaging to go. It goes into the dump, where it is usually burned so bears aren't attracted to town, and then the scraps blow into the water. There is no way to behave differently. Bag bans don't eliminate the problem. Degradable plastics made of corn would move the problem onto someone else's land. Shipping Nain's plastics to a recycling plant in Vietnam or even elsewhere in Canada produces pollution and plastic leakage on other lands still. Disposable plastics are simply not possible without colonizer access to land. The end of colonialism will result in the end of plastic disposability.

# Climate Disaster Is a Labor Issue: Here's Why

### KIM KELLY

*May 23, 2019*

The climate crisis is the greatest threat facing humanity. Given the number of imperialistic wars, white supremacist terrorist attacks, mass extinctions, concentration camps, genocides, and brutal government repression with which we as a species are currently occupying ourselves, that is truly saying something.

A widely cited report from the United Nations Intergovernmental Panel on Climate Change, released in 2018, warned that, unless a massive effort is undertaken to curtail the climate emergency, Earth has only a few short years before it enters an irreversibly apocalyptic scenario. According to a 2015 World Bank report, the planet only has until 2030 before the climate crisis forces 100 million more people into poverty.

The effects of climate disaster are already here: cyclones, tsunamis, wildfires, toxic air, disappearing islands. These increasingly common events have had a disproportionate effect on poor people, who have less access to crucial infrastructure and resources when catastrophe strikes; these events will continue to hit the hardest those living in poverty, while the rich continue to profit off the suffering of the poor.

For this and many other reasons, the breakdown of our climate is not only a looming peril—it's a labor issue.

The Green New Deal, a proposal introduced by Democratic lawmakers Representative Alexandria Ocasio-Cortez and Sen. Ed Markey and galvanized by an engaged youth movement, is perhaps the U.S. government's most robust attempt to make a dent in the country's outsized carbon emissions and end its reliance on fossil fuels, both of which propel the ongoing climate disaster when in use (burning them releases harmful chemicals into the atmosphere) and during extraction (via methods like oil drilling and coal mining, which cause irreparable damage to the environment). The Green New Deal also seeks to address the dire income inequality that has existed since European colonizers stepped foot on this native land, and has only been exacerbated by climate change, both here and on global terms. This policy proposal isn't a fix-all, but it is an ambitious program that, if implemented, has the potential to enact real, much-needed change, especially if other entities (for example, New York City, whose city council recently passed a Climate Mobilization Act) are inspired to take action on a local level.

At a glance, the proposal seems extremely union-friendly. A core component of the package focuses on workers like coal miners and oil riggers, whose jobs have been and will continue to be impacted by a national pivot away from extractive industries, and promises to create "high-quality union jobs" while protecting the rights of workers to organize and collectively bargain. The program also seeks to guarantee health care, housing, and a job with "a family-sustaining wage, adequate family and medical leave, paid vacations, and retirement security to all people of the United States"; an enticing proposal in a nation where approximately thirteen million children are living under the poverty line. But the Green New Deal has gotten a surprising amount of pushback from certain sectors of the labor movement. In March, the American Federation of Labor and Congress of Industrial Organizations' (AFL-CIO) energy committee released a statement against the proposal, written on behalf of a number of manufacturing unions like the United Steelworkers and United Mine Workers of America (UMWA), which said the provisions it made for potentially impacted workers were not detailed enough, arguing that its goal to transition away from fossil fuels would immediately kill off jobs. Other labor leaders are skeptical of the proposal's promises to transition workers and "green jobs" guarantees. As Phil Smith, a spokesman for UMWA, told Reuters, "We've heard words like 'just transition' before, but what does that really mean? Our members are worried about putting food on the table."

That some workers are feeling left behind and have already witnessed firsthand how institutional neglect can ravage working-class communities is a familiar concept to me. When the then New Jersey governor Chris Christie approved a new natural gas pipeline project that would snake through the pristine wilderness of the Pine Barrens, the South Jersey nature preserve where I grew up and my entire family still lives, my first thought was of my dad: he's been a union construction worker since he was eighteen, and chances were he'd be called to work on the pipeline, if it ever made it out of the courts.

The thought of him digging up his beloved Pinelands broke my heart, but he remained stoic at the prospect. After all, he needed the money to cover bills and care for my disabled mother; as much as he loves those woods, his hands would be tied.

Make no mistake: the coal miner and pipeline worker know about the environmental costs of their labor, but when faced with the choice of feeding their kids or putting down their tools in the name of saving the planet, the pressures of capitalism tend to win; their choice is made for them. That is why it's so important to dismantle the structures that force these impossible decisions and offer instead real, equitable alternatives to those whose livelihoods depend on industries that harm the earth.

While many in the labor movement were disappointed in the AFL-CIO energy committee's stance, and many individual unions have endorsed the Green New Deal, it's not hard to see why some union members in certain industries would be wary of any big shifts. Job-retraining programs have had mixed success in areas like Appalachia, where coal miners have found new lines of work, thanks to grassroots community efforts, but have also been the victims of botched schemes like "coding boot camps" run by clueless outsiders. Studies have shown that retraining miners to work in the renewable energy industry would be both cost-effective and financially viable for the workers themselves, but efforts like these require institutional support (and funding) to really make an impact. Theoretically, the Green New Deal would supply said resources, but what if we didn't have to wait for the government to finally, fitfully pass an inevitably watered-down version of someone else's vision?

What if, instead, the labor movement took matters into its own hands, and we seized control of our future?

Here's the thing: the AFL-CIO, the largest federation of unions in the U.S., has a lot of money. In 2018, the organization spent over $4 million in donations to various politicians and lobbying groups. I'm not proposing that it drop everything and pour its resources into job-training programs, but what if the organization launched its own large-scale, nationwide program offering job training specifically to the workers whose continuing careers the energy committee was justifiably concerned about in its statement?

The union can't take on this entire burden alone—that's the kind of heavy lift that only the government has billions of dollars available to address in a meaningful way. However, if the AFL-CIO threw its weight—financially, but more importantly, politically—behind a government-sponsored job-training program, like the one proposed in the Green New Deal, it would make a huge difference. And on their own terms.

The AFL-CIO already offers a robust number of apprenticeship programs in manufacturing, building, and construction trades, and represents over twelve million workers in fifty-five national and international labor unions. That's a lot of people power, and a lot of opportunities to act. As a member of an AFL-CIO affiliate, the Writers Guild of America, East, I would be thrilled to see my union dues help fund an expansion and retooling of programs aimed at preparing fellow workers in the oil, gas, and coal industries for new employment in the renewable energy sector (or other sectors—there's a dire shortage of home health care workers, for example).

Some International Brotherhood of Electrical Workers (IBEW) locals have already started experimenting with these kinds of programs in California, and the AFL-CIO itself has a long-running relationship with the environmentally focused Sierra Club. Even AFL-CIO president Richard Trumka, a former coal miner, has spoken about the need to take "bold, comprehensive action to fight climate change," and advocated for funding investments in technology to help workers build a more sustainable economy.

Few labor leaders deny that some kind of action needs to be taken, but it's time to pick up the pace.

The framework is there, should we choose to use it. By funneling resources to local union affiliates in coal- and oil-producing areas like Appalachia, Texas, North Dakota, Alaska, and New Mexico, and engaging with frontline communities, the AFL-CIO could get ahead of whatever's coming next—whether that's the passage of the Green New Deal, or, perhaps more likely, the continuing death of the coal industry. The union leaders were right to worry about the future, but their vision is too clouded by oil to see the burning forest.

Others in the movement have already stepped up. The Labor Network for Sustainability was launched by labor veterans (including former AFL-CIO employees) to support workers and communities in building "a just transition to a climate-safe and equitable economy"; its current project, Making a Living on a Living Planet, seeks to strengthen the relationship between labor and environmentalism.

The BlueGreen Alliance, a group of labor unions and environmental organizations, advocates for "clean jobs, clean infrastructure, and fair trade" via research, public policy, advocacy campaigns, and education. Its membership combines union heavyweights like the United Steelworkers, Service Employees International Union (SEIU), United Association of Plumbers and Pipefitters (UAPP), and the American Federation of Teachers, with venerated environmental institutions like the National Wildlife Federation, making it clear that many unions want to make climate change action a priority.

Association of Flight Attendants-Communication Workers of America president Sara Nelson's own industry is grappling with both the causes and effects of climate change. In a recent interview with *In These Times*, she addressed the need to build support for the Green New Deal, and renewable energy in general, emphasizing the need for a holistic approach. She said:

> We must recognize that labor unions were among the first to fight for the environment, because it was our workspaces that had pollutants, our communities that industry polluted. . . .
>
> We need to build a broad coalition. And to do that we can't start from a position that assumes opposition. If we bring everyone to the table, recognize the efforts to date, draw on the expertise from each affected field, and mobilize a united effort, then we can create allies where we otherwise might have had enemies.

At this point, those who are determined to take action to address the climate crisis need all the allies they can get. Time is running out for all of us, but at a faster pace for the poor and working class. Now, as ever, any real change is going to begin at the grassroots level, but labor leaders also need to take concrete steps toward alleviating the chaos to come by investing in workers' futures, instead of wasting time boosting spineless politicians. Right now, our future is in flames, and some of those politicians are standing on the hose.

# Four Activists Explain Why Migrant Justice Is Climate Justice

## MAIA WIKLER

*June 11, 2019*

The climate crisis, fueled by capitalism, colonialism, and imperialism, is making many parts of the world inhospitable. Because of this crisis, the climate-fueled movement of people is already well underway.

Climate disaster is fueling more-frequent droughts, flash floods, and food shortages; causing dwindling water supplies; and impacting land that people rely upon. Migration is happening where homelands become inhabitable, often for those with the least amount of resources to adapt to climate change.

In 2018, the fourth-hottest year on record, we saw increases in record-setting wildfires in North America, devastating hurricanes, flash floods in India, a typhoon in the Philippines, and deadly wildfires in Greece and Sweden. And the Arctic experienced its second-warmest year on record, with a five-year heat streak, warming at a rate twice as fast as the rest of the world.

A U.N. Refugee Agency report revealed that, by the end of 2016, there were 65.6 million displaced people who had fled their homelands because of violence, human rights violations, and environmental disasters that are intensified by the climate crisis. Since 2008, an average of 26.4 million people have been displaced from their homes by extreme weather disasters every year. 350.org stated in a December press release:

> From African migrants choosing to cross by boat from North Africa to Europe to Pacific Islanders losing their homes due to rising sea levels and Central American migrants fleeing their home countries in search of refuge, people around the world are being driven from their homes by droughts, storms, and the political strife and conflict that follow these climate disasters.

Fighting climate change is about more than emissions and metrics—it's about fighting for a just world for everyone. *Teen Vogue* spoke with five climate-justice advocates whose work focuses on the vital intersection of migrant rights and climate action.

Maya Menezes, an organizer for No One Is Illegal and podcast host of *Change Everything*:

> We are past the point of stopping some of the largest impacts of climate change. One of the biggest battles will be over the closing of borders, the decisions of who is deserving of basic humanity and who isn't.
>
> Under capitalism, goods can go across borders but human beings cannot. It's not a weird coincidence, it's a violent political strategy to bar people and privilege some over others. We need to envision a borderless world. Imagining a borderless world is one

of the ultimate acts of decolonization because colonialism told us arbitrarily there are lines here for you to cross; it is connected to capitalism, exploitation, and racism, so challenging capitalism and colonization fundamentally challenges borders. If we are trying to challenge capitalistic structures that are destroying this planet, that means challenging the structures that are continuing to dehumanize human beings and designating people as legal bodies. No one is illegal on stolen lands. If we reject colonization and put ourselves in solidarity with Indigenous sovereignty, then we reject that someone can be illegal and discarded.

Getting involved in climate justice work involves everything, it's not as simple as recycling or buying local. It's everything from deciding not to be a border enforcer in your community, to being in solidarity with complex Indigenous movements all over the world. Capitalism individualizes our suffering. Moving away from individualizing hardship and instead collectivizing our struggles is an empowering act. Go out into your communities and join collectives, collective movements are the way we fight individualism and capitalism—that we are in this together as opposed to doing this on our own.

Nayeli Jimenez, youth organizer for Our Time 2019 and art director at Greystone Books:

I'm from Cuernavaca, Mexico, and in my country, the desire to go to another place comes from needing to have a safer place to live. It's really heartbreaking to leave the country you grew up in, to know that my family is so far away and also happier knowing I'm safe, because where I'm from is so wrought with violence against women and the land. The main reason why people leave is so they can keep living. People are being met with resistance and walls.

They don't want to leave their homes. If it were up to us, our communities back home would be thriving and we wouldn't be forced out because of violence or extreme weather conditions. Closed borders are a violation of human rights.

I now live in Canada, and I can advocate for Mexico because I'm at a distance, which gives me safety. Living in deep colonial oppression, you have no say on your life, and you are being forced out of your homelands. The process of immigration is so dehumanizing—having gone through it, I know firsthand. We need systems in place to make sure people coming here to seek asylum have access to opportunities to thrive, rather than being othered.

**Niria Alicia, Xicana community organizer and SustainUS COP25 youth delegation leader:**

I am Xicana, first-generation born and raised in southern Oregon in a migrant farmworker community. I [worked] as a young kid myself as a farmworker, [and] both of my parents were migrants. I grew up in a mixed-status community that kept me deeply in a relationship to the land and the environmental changes. As a kid, I always heard my mom talk about climate change from the observations of having to plant seeds that we brought from Mexico. Especially farming pears, you are always watching temperature drops so the blossoms don't frost. In a migrant farmworker community, those two connections of climate change and migrant rights are so present. We are directly experiencing these changes, being exposed to industrial agriculture and chemicals and living in terror of what will happen if we get pulled over [while undocumented].

This culture of disposability and treating the earth as just a resource is the same way that capitalism treats migrant workers as disposable and not worthy of being protected from cancerous chemicals. The climate-justice [movement] needs to understand that if we aren't in solidarity with refugees, migrants, and people displaced by climate change, war, and violence, we are doing ourselves a disservice, because we will need to deal with these issues soon, whether we like it or not.

I volunteer with No More Deaths, which is an organization in Arizona that works to end unnecessary death and suffering of human beings on the U.S.-Mexico border. I just finished a month-long program with volunteers there doing drops of water, food, blankets, socks, any necessary supplies migrants need to make it out of this journey alive. Going out in the winter, the desert is cold. During my time there it snowed; we were going out there doing drops in the snow, we were finding people's shoes, jackets, rosaries. It absolutely broke my heart to see that, despite the weather conditions, people are still moving through these trails. How bad does it have to get for someone to leave the comfort of their home, to migrate through rough terrain in a country where they don't speak the language in hopes of surviving? A lot of these people are out there for days, weeks, months, being hunted by border patrol.

We haven't figured out how to keep humanity legal. We need to reconnect ourselves; we need to learn to look at people and land and water as a spirit and entity that is invaluable, that cannot be commodified and monetized. When we do that it will shift the way we relate to the earth and ourselves.

Thanu Yakupitiyage, Sri Lankan–born, Thailand-raised, Brooklyn-based multidisciplinary artist and associate director of U.S. communications at 350.org:

In most places, immigration policy doesn't account for displacement based on the climate crisis. In the U.S., for example, you can't apply for refugee status because of climate impacts, [though] one of the factors driving migration from Central America is drought and food insecurity. There's no international law or national refugee law that says climate impacts are a reason to allow asylum or refugee status.

We are in a severe right-wing era, from the U.S. to Brazil and India to places in Europe. If we don't talk about climate justice as human rights, we aren't going to be able to move forward [with] solutions, particularly for communities from the Global South. What are we going to do about the reality of hundreds of thousands of people moving? We need to be able to support folks fighting for the survival of their land and the right to move.

What does global migration reform look like so we can support people impacted by displacement? There is so much more we could be doing with the Green New Deal—are folks who are advocating for the Green New Deal also going to call for those jobs to be supportive of migrants? What do climate reparations look like? Where can the Green New Deal intersect with refugee and asylum reform to allow for climate refugees to enter the U.S.? A radical manifestation of how we think about immigrant rights in the U.S. has been the Abolish ICE movement. I personally think we absolutely need to abolish ICE and reconsider the current system of borders. If we start at a place of radical imagination and re-envisioning, then we can think about [what] truly will work [to] uplift all communities.

# Alexandria Ocasio-Cortez Said Climate Change and Immigration Are Connected— Here's Why She's Right

## LUCY DIAVOLO

*April 11, 2019*

Representative Alexandria Ocasio-Cortez (D-NY) had a double-trouble callout on April 9 for people on the far right. The congress-woman known as AOC has been a leader on the Green New Deal, a plan to tackle both climate change and economic inequity. But on Tuesday, she pointed out how efforts to ignore climate change and stoke fears about immigration are connected.

"The far-right loves to drum up fear and resistance to immigrants. But have you ever noticed they never talk about what's causing people to flee their homes in the first place?" she wrote on Twitter. "Perhaps that's [because] they'd be forced to confront one major factor fueling global migration: climate change."

AOC is right. The environmental impacts of climate change and disaster-level events related to it are displacing people from their homes all across the planet. Many of these people are fleeing destroyed homes and livelihoods in the wake of climate change–related crises. But at the same time, it's getting harder to seek refuge.

The necessity of migration when climate change hits home is all too real around the world. According to the Science and Development Network, people in Bangladesh have been forced to flee due to rising sea levels brought on by climate change. According to the European Council on Foreign Relations, it's believed that a lack of protective infrastructure and a reliance on economic sectors vulnerable to climate change impacts migration in Africa. And Public Radio International reported last year that climate change–related droughts and other natural disasters are creating refugees from Central America, a region with migrants President Donald Trump is fond of attacking.

For those fleeing the devastation of climate change, the tricky business of crossing international borders complicates things in an era when leaders in the United States and several European countries are cracking down on immigration, even though U.S. companies like ExxonMobil and Chevron and European companies like Shell and BP have historically been among the highest polluters.

AOC's tweet was issued in response to a video from a group called The Leap, an environmental justice organization that says it "makes system change irresistible." Narrated by actress (and *Teen Vogue* cover star Nico Parker's mom) Thandie Newton, the video breaks down how right-wing fear-mongering over immigration connects to right-wing denial of climate science.

"From wildfires in Alberta to hurricanes in Puerto Rico, climate change is one of the reasons many of us are forced to leave our homes in search of a safer place to live," Newton says in the video. "We keep hearing that migration is a crisis, and it is, for the people affected. But did you ever notice that the same leaders denying climate change are the ones drumming up fear and hatred against migrants?"

"To win climate justice, we need to oppose racism," Newton says. "We know the world's wealthiest countries have burned most of the carbon that is driving climate change today. Asserting the rights of migrants affected by these storms, floods, and fires is a way of paying back our climate debts."

The way the environment and race intersect isn't just something that occurs outside our borders. In the United States, the Flint water crisis and the Dakota Access Pipeline are examples of a concept called "environmental racism," the idea that negative environmental impacts are more likely to affect people of color. Puerto Rico is another prime example of environmental racism in a U.S. territory, as Hurricane Maria created thousands of climate refugees and the federal government reportedly provided less funding, and did it more slowly, in Puerto Rico than after similar-strength hurricanes in Texas and Florida.

Indigenous peoples all over the world are grappling with climate issues firsthand. And in places like Malawi and Bangladesh, the impacts of climate change are often gendered, impacting women and girls of color more severely than anyone else.

Just like The Leap group, young activists at climate marches around the world have expressed an interest in systemic change. System Change Not Climate Change, an anti-capitalist, eco-socialist activist network, is one example of a group that lays out how a massive overhaul of current economics and politics may be necessary to address climate change, and questions whether or not those changes are incompatible with capitalism.

Whatever the overhaul looks like, it must include making space for people fleeing from climate-related disasters brought on by the emissions produced by the planet's worst polluting companies. Acknowledging the intersections between climate and migration is essential to tackle right-wing fear tactics and ensure that the drastic change needed can happen.

# Acknowledgments

I regularly tell everyone that the political coverage from *Teen Vogue* can't be attributed to a single person but has taken at least a village if not an entire city. This book is no different.

As its editor, I'd like to thank the following people from *Teen Vogue*: Lindsay Peoples Wagner and Samhita Mukhopadhyay for their leadership, Allegra Kirkland for tagging into our team, Alli Maloney for getting my foot in the door, Phillip Picardi for giving me a chance, Ella Cerón for showing me the ropes, my coworkers for making the office not just bearable but also fun, and every writer who's ever graced us with the privilege of publishing their work.

I'd also like to thank everyone at Haymarket Books who helped make this project a reality, including Anthony Arnove for going to the mat, Dana Blanchard for her eminently personable patience and persistence, and Maya Marshall for using a fine-toothed comb on our copy.

I'd of course like to thank my parents for instilling in me an environmental sensibility from my earliest days, my sister for being an inspiration, and all the friends and loved ones who have been a wind in my sails on stagnant seas or a hand reaching out to help me back to my feet.

Finally, I'd like to thank the young people who read our work and who allow us to share their stories. You may not sign my paychecks, but my work at *Teen Vogue* has always been for you.

Lucy Diavolo
May 2020

# Glossary

**American Federation of Labor and Congress of Industrial Organizations (AFL-CIO)**: the democratic, voluntary federation of fifty-five national and international labor unions that represent 12.5 million working men and women. It is the largest federation of unions in the United States.

**Black lung disease**: a common name for any lung disease that develops from inhaling coal dust. This name comes from the fact that those with the disease have lungs that look black instead of pink. Medically, it is a type of pneumoconiosis called coal workers' pneumoconiosis (CWP).

**Climate change**: a term that describes changes in average weather conditions that persist over multiple decades. Climate change encompasses both increases and decreases in temperature, as well as shifts in precipitation; changing risk of certain types of severe weather events, and changes to other features of the climate system.

**Climate refugees**: also known as climate migrants; a subset of environmental migrants who were forced to flee due to sudden or gradual alterations in the natural environment related to at least one of three impacts of climate change: sea level rise, extreme weather events, and drought and water scarcity.

**Clean Power Plan**: announced by President Obama in August 2015, the plan set the first-ever limits on carbon pollution from U.S. power plants, the largest source of the pollution in the country that's driving dangerous climate change.

**Colonialism:** the policy or practice of acquiring full or partial political control over another country, occupying it with settlers, and exploiting it economically. In this process, the colonizer imposes their religion, economics, and other cultural practices on occupied peoples.

**Environmental classism**: the disproportionate impact of environmental hazards on low-income people and neighborhoods. This occurs when poor neighborhoods, towns, and cities are unjustly subjected to hazardous surroundings in a manner that wealthier areas aren't.

**Environmental Protection Agency**: an independent executive agency of the United States federal government for environmental protection. Its stated mission is to protect human health and the environment. More information is available at epa.gov.

**Environmental justice:** a movement aimed at addressing and abolishing environmental racism and environmental classism. The EPA defines environmental justice as "the fair treatment and meaningful involvement of all people regardless of race, color, national origin, or

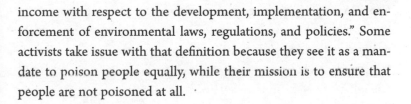

income with respect to the development, implementation, and enforcement of environmental laws, regulations, and policies." Some activists take issue with that definition because they see it as a mandate to poison people equally, while their mission is to ensure that people are not poisoned at all.

**Environmental racism**: the disproportionate impact of environmental hazards on people of color. This occurs when corporations or local, state, and federal governments target and unfairly subject minority communities to unhealthy living conditions.

**Extinction Rebellion**: a global environmental movement with the stated aim of using nonviolent civil disobedience to compel government action to avoid tipping points in the climate system, biodiversity loss, and the risk of social and ecological collapse. More information is available at https://rebellion.earth.

**Food desert**: an area that lacks access to affordable and nutritious foods.

**Fossil fuels**: fuel formed by natural processes, such as anaerobic decomposition of buried dead organisms, containing organic molecules originating in ancient photosynthesis that release energy in combustion.

**Fracking:** the process of injecting liquid at high pressure into subterranean rocks, boreholes, and other formations. so as to force open existing fissures and extract oil or gas.

**Global warming**: the ongoing rise of the average temperature of the earth's climate system. It is a major aspect of climate change which, in addition to rising global surface temperatures, also includes its effects, such as changes in precipitation.

**Global Youth Climate Strike**: the September 2019 climate strike, also known as the Global Week for Future, consisted of a series of international strikes and protests to demand action be taken to address climate change; it took place September 20–27.

**Greenhouse effect**: the process by which radiation from a planet's atmosphere warms the planet's surface to a temperature above what it would be without this atmosphere. Radiatively active gases in a planet's atmosphere radiate energy in all directions.

**Greenhouse gases**: Greenhouse gases that absorb and emit radiant energy within the thermal infrared range. Greenhouse gases cause the greenhouse effect on planets. The primary greenhouse gases in Earth's atmosphere are water vapor, carbon dioxide, methane, nitrous oxide, and ozone.

**Green New Deal**:  a proposed package of United States legislation introduced by Representative Alexandria Ocasio-Cortez of New York and Senator Edward J. Markey of Massachusetts that aims to wean the United States from fossil fuels. It also aims to guarantee new high-paying jobs in clean-energy industries, and to address climate change and economic inequality. The name refers back to the New Deal, a set of social and economic reforms and public works projects undertaken by President Franklin D. Roosevelt in response to the Great Depression.

**Indigenous:** describes First peoples, First Nations, Aboriginal peoples, or Native peoples; ethnic groups who are the original or earliest known inhabitants of an area prior to colonization.

**Microplastics:** not a specific kind of plastic but rather any type of plastic fragment that is less than 5 mm in length according to the U.S. National Oceanic and Atmospheric Administration (NOAA). They enter natural ecosystems from a variety of sources, including cosmetics, clothing, and industrial processes.

**The Paris Agreement:** a landmark environmental accord that was adopted by nearly every nation in 2015 to address climate change and its negative impacts. The deal aims to substantially reduce global greenhouse gas emissions in an effort to limit the global temperature increase in this century to 2 degrees Celsius above preindustrial levels, while pursuing means to limit the increase to 1.5 degrees. The agreement includes commitments from all major emitting countries to cut their climate-altering pollution and to strengthen those commitments over time. The pact provides a pathway for developed nations to assist developing nations in their climate mitigation and adaptation efforts, and it creates a framework for the transparent monitoring, reporting, and ratcheting up of countries' individual and collective climate goals.

**Petrochemicals:** the chemical products obtained from petroleum by refining. Some chemical compounds made from petroleum are also obtained from other fossil fuels, such as coal or natural gas, or renewable sources such as maize, palm fruit, or sugar cane.

**Pro-environmental behavior**: a set of behaviors that a person consciously chooses in order to minimize the negative impact of their actions on the environment.

**The Red Deal**: drawing from Black abolitionist traditions, this platform calls for divestment away from the criminalizing, caging, and harming of human beings *and* divestment away from the exploitative and extractive violence of fossil fuels. It was crafted by community members, Native people, young people, and poor people and has four key tenets designed to build on and push forward the ideas in the Green New Deal. First, what creates crisis cannot solve it; second, change must come from below and move to the left; third, politicians can't do what mass movements do; and fourth, the climate conversation must move from theory to action.

**The three R's**: reduce, reuse, and recycle—all help to cut down on the amount of waste we throw away. They conserve natural resources, landfill space, and energy.

**Salinization:** the process by which water-soluble salts accumulate in the soil. It is a resource concern because excess salts hinder the growth of crops by limiting their ability to take up water. The process may occur naturally or because of conditions resulting from management practices.

**Slash-and-burn agriculture**: a widely used method of growing food in which wild or forested land is clear cut and any remaining vegetation burned. The resulting layer of ash provides the newly cleared land with a nutrient-rich layer to help fertilize crops. However, under this method, land is only fertile for a couple of years before the nutrients are used up. Farmers must abandon the land, now degraded, and move to a new plot—clearing more forest in order to do so.

**Sunrise Movement**: an American youth-led political movement coordinated by Sunrise, a 501 political action organization that advocates political action on climate change.

**Two spirit**: a term coined in 1990 as a means of unifying various gender identities and expressions of Native American/First Nations/Indigenous individuals. The term is not a specific definition of gender, sexual orientation, or other self-determining catchall phrase but rather an umbrella term. The term does not diminish the tribal-specific names, roles, and traditions nations have for their own two-spirit people. Examples of such names are the *winkte* among the Lakota and the *nadleeh* among the Navajo people.

**Waste colonialism**: a term describing how waste and pollution are part of the domination of one group in their homeland by another group.

**Zero Hour**: a nonprofit that seeks to center the voices of diverse youth in the conversation around climate and environmental justice. Zero Hour is a youth-led movement creating entry points, training, and resources for new young activists and organizers (and adults who support its vision) wanting to take concrete action around climate change. Together, the participants are a movement of unstoppable youth organizing to protect young people's rights and access to the natural resources and a clean, safe, and healthy environment that will ensure a livable future where they not just survive, but flourish. More information is available at http://thisiszerohour.org.

# Index

"Passim" (literally "scattered") indicates intermittent discussion of a topic over a cluster of pages.

No Planet B

# Contributor Biographies

**Sarah Emily Baum** is a multimedia journalist. She was a senior reporter for the multiple award–winning "Since Parkland" project with the *Trace* and the *Miami Herald*.

**Roxanna Pearl Beebe-Center** is a teen reporter and activist.

**Lincoln Anthony Blades** is a journalist, a podcaster, and a digital media director/digital strategist who has reported on and critiqued matters of race and culture, criminal justice reform, and geopolitical conflicts.

**Rosalie Chan** is a senior reporter the *Business Insider* covering enterprise tech. She writes a newsletter on stories by women of color: truecolors.substack.com

**Michael Charles** is an activist and speaker from the Navajo Nation.

**Lucy Diavolo** is a news and politics news editor for *Teen Vogue*. She helped found the Transfeminine Alliance of Chicago and facilitates its regular meetings. She also plays bass in the Chicago-based band The Just Luckies.

**Denise Garcia** is a writer, artist, and performer. Her writing can be found at *Teen Vogue* Culturebanx and CNBC.

**Mélissa Godin** is a journalist reporting on climate change, gender, public health, and human rights. She is currently writing for *Time* magazine's International Desk in London and has previously worked for the *New York Times* Paris bureau.

**Isabella Gomez Sarmiento** is a 2019 Kroc Fellow reporting for *Goats and Soda*, the National Desk, and NPR's *Weekend Edition*. She has published with *Teen Vogue*, CNN, and Remezcla.

**Destine Grigsby** is a teen activist and writer.

**Emily Hernandez** is a writer who has published articles for *Teen Vogue* on climate change, current environmental events, politics, national monuments, and minorities' access to public lands.

**Ruth Hopkins** is a Dakota/Lakota Sioux writer and enrolled member of the Sisseton Wahpeton Sioux Tribe. She is also a biologist, tribal attorney, former judge, and cofounder of Lastrealindians.com. Ruth resides on the Lake Traverse Reservation in South Dakota.

**Kareeda Kabir** is a media and communications professional pursuing a bachelor of arts in linguistics.

**Kim Kelly** is a freelance journalist and anarchist organizer based in Philadelphia. Follow her on Twitter at @grimkim.

**Allegra Kirkland** is senior politics editor for *Teen Vogue*.

**Marilyn La Jeunesse** is a contributing writer for INSIDER. She is a freelance social media strategist and writer with words in *Teen Vogue*, *Mashable*, *Mic*, and *Domino*.

**Ray Levy-Uyeda** is a Bay Area–based freelance writer who focuses on gender, politics, and activism. You can find her work elsewhere at *Teen Vogue*, *Fortune*, and *Vice*. Find her on Twitter at @raylevyuyeda.

**Dr. Max Liboiron** is an artist who also directs the Civic Laboratory for Environmental Action Research (CLEAR) at Memorial University of Newfoundland.

**Alli Maloney** is a freelance writer and a news and politics editor for *Teen Vogue*.

**Greta L. Moran** is a writer living in Queens, New York. Moran has written for *Atlantic*, *Guardian*, *New Republic*, *New Yorker* online, and elsewhere. Moran writes narrative-based stories about public health, climate change, and environmental justice, with a lens on the people working towards solutions.

**Samhita Mukhopadhyay** is the executive editor of *Teen Vogue* and the coeditor of the anthology *Nasty Women: Feminism, Resistance, and Revolution in Trump's America*.

**Lindsay Peoples Wagner** is the editor in chief of *Teen Vogue*.

**Maia Wikler** is an activist, investigative researcher, and writer. Wikler seeks to connect dynamic and diverse audiences on issues of climate justice and forced displacements via academia, film, writing, and community organizing.